# Natural Disasters and Development in a Globalizing World



The number of humanitarian disasters triggered by a natural hazard has doubled every decade since the 1960s. At the same time, the global economic growth rate per capita is twice its 1960s value. Does this mean economic growth is independent of the impacts of natural disaster? This was certainly the view for many years, but the scale of human lives and economic assets at risk today means that the global economy itself is at risk from catastrophe.

As we become aware of the global scale processes of environmental change and economic liberalization, it is becoming increasingly clear how fundamental these global pressures are for shaping local geographies of risk. The contributors to this book look at the disaster–development relationship under globalization, from three different perspectives. First, there is an examination of global processes and how they might affect disaster risk at the global scale. Second, links between international issues, such as diplomatic relations, the growth of non-governmental organizations and the health of the international insurance industry, and disaster risk are explored. Third, the interaction of these large-scale forces with local conditions are examined through case study analysis of individual disaster events, from the so-called developed and developing worlds.

*Natural Disasters and Development* makes clear that there are links between global scale processes and local experiences of disaster, but underlines the difficulty of attributing blame for individual disasters on specific global pressures. It is argued that action to reduce disaster needs to be co-ordinated at the local, national and global scales and that there is a need for greater integration across the physical and social sciences. In this context, the human rights agenda is seen as a way of moving disaster reduction efforts forward.

Edited by **Mark Pelling**, Lecturer in Geography at the University of Liverpool.

# Natural Disasters and Development in a Globalizing World

## Edited by Mark Pelling

Routledge
Taylor & Francis Group
LONDON AND NEW YORK

First published 2003
by Routledge
11 New Fetter Lane, London EC4P 4EE

Simultaneously published in the USA and Canada
by Routledge
29 West 35th Street, New York, NY 10001

*Routledge is an imprint of the Taylor & Francis Group*



Typeset in 10/12pt Times New Roman by Graphicraft Limited, Hong Kong
Printed and bound in Great Britain by MPG Books Ltd, Bodmin



*British Library Cataloguing in Publication Data*
A catalogue record for this book is available
from the British Library

*Library of Congress Cataloging in Publication Data*
Natural disasters and development : in a globalizing world / edited by
Mark Pelling.
p. cm.
Includes bibliographical references (p. ).
1. Economic development. 2. Natural disasters.
3. Globalization. I.
Pelling, Mark, 1967–
HD75 .N396 2003
363.34′2--dc21 2002011567

ISBN 0-415-27957-7 (hbk)
ISBN 0-415-27958-5 (pbk)

# Contents

# Figures

# Tables

# Boxes

# Contributors

**W. Neil Adger** is Reader in the School of Environmental Sciences, University of East Anglia, Norwich, UK. He is a Senior Research Fellow in the CSERGE and manages the programme on Adaptation for the Tyndall Centre for Climate Change Research. His research has focused on climate change adaptation, vulnerability and resilience. Recent co-authored books include *Living with Environmental Change: Social Vulnerability, Adaptation and Resilience in Vietnam* (Routledge 2001), *The Economics of Water and Coastal Resources* (Kluwer 2001) and *Making Waves: Integrating Coastal Conservation and Development* (Earthscan 2002). Contact N.Adger@uea.ac.uk

**Katrina Allen** is a researcher at the Flood Hazard Research Centre, University of Middlesex, UK. Most recently, she has conducted work on community-based rural development projects in Ethiopia for FARM Africa and the Gudina Tumsa Foundation, and community based approaches to disaster mitigation supported by the International Federation of the Red Cross and Red Crescent. Contact K.Allen@mdx.ac.uk

**Nick Brooks** is Senior Research Associate in CSERGE and the Tyndall Centre for Climate Change Research. He has a background in climate change and earth sciences, with a special interest in climatic, environmental and social change in northern Africa. He was awarded a Ph.D. by the Climatic Research Unit in 2001 (on climate change and dust production in the Sahel and Sahara), and has participated in environmental and archaeological fieldwork in the Sahara. He is currently working on a project assessing the vulnerability of different countries to climate change. Contact n.brooks@uea.ac.uk

**Ian Christoplos** is an independent researcher and consultant working with issues related to humanitarianism, risk and rural development. His interests focus on policy formation for poverty reduction and the supporting role of local institutions in turbulent contexts. He has worked as a researcher and practitioner in both development and humanitarian assistance in Africa, Asia and Latin America. Contact christop@bahnhof.se

**Mohammed H.I. Dore** is Professor of Economics, Brock University, Canada. He has published numerous book chapters and papers and five books on the themes of economic development, environmental economics and global climate change. He was recently appointed by UNESCO to be an editor of one of the volumes of the UNESCO *Encyclopedia of Life Support Systems.* Contact dore@adam.econ.brocku.ca

**David Etkin** is Adjunct Professor at the University of Toronto and leads the Canadian Natural Hazards Assessment Project. He has an M.Sc. in Physics and is a scientist with the Adaptation Impacts Research Group at Environment Canada. His recent research interests include tornado and hail hazard in Canada; natural hazard (mainly atmospheric) risk assessment, incorporation of climate change into environmental assessments and into design criteria. Contact david.etkin@ec.gc.ca

**Maureen Fordham** is Head of the Department of Geography and Senior Lecturer at Anglia Polytechnic University, Cambridge, UK and First Vice-President of the International Sociological Association's Research Committee on Disasters. While gender and disasters is her main research area, she also has a strong interest in environmental management. Recent work includes contributing to the United Nations Expert Group Meeting on Environmental Management and the Mitigation of Natural Disasters. She was co-editor of *Floods Across Europe: Flood Hazard Assessment, Modelling and Management*. Contact m.h.fordham@anglia.ac.uk

**Chen Guojie** is Professor at the Chengdu Institute of Mountain Hazards and Environment, Chinese Academy of Sciences, Chengdu, People's Republic of China. His research interests include environment and regional sustainable development. Contact klpcgj@mail.sc.cninfo.net

**Jacqueline Homan** is Lecturer in Human Geography at the University of Wolverhampton. Her research has involved exploring how both technological and environmental risks are understood by the public and how communication of these risks could be improved to reduce exposure. She has researched earthquakes in both Egypt and Turkey and, more recently, has undertaken work in the UK for government departments and public agencies such as the Department of Health and the Environment Agency. Contact J.Homan@wlv.ac.uk

**Ilan Kelman** studied at the University of Toronto and the University of Cambridge. His academic background is in engineering, but his research focus has been on disasters, vulnerability, development and environmental management. His main geographical interest is small islands and his publications include work on volcanoes, floods and emergency shelter. Contact ik227@hermes.cam.ac.uk

**Arthur Morris** is Emeritus Professor at the University of Glasgow, Department of Geography. His main research focus is in South America, on themes of regional development and development agencies. In the Central Andes, his work includes environmental aspects of development projects. His most recent book is *Geography and Development* (UCL Press). Contact amorris@geog.gla.ac.uk

**Alpaslan Özerdem** is Lecturer in Politics at the University of York and visiting lecturer at the universities of Birmingham and London. He is active in research and consultancy in post-conflict recovery, reconstruction of infrastructure, reintegration of former combatants and mitigation strategies for earthquake disasters. Contact ao102@york.ac.uk

**Mark Pelling** is Lecturer in Geography at the University of Liverpool. His work addresses environmental risk, adaptation to global environmental change and the sociology of international development. He has published widely from research funded by the Economic and Social Research Council and the British Academy, his most recent book being *The Vulnerability of Cities: Social Adaptation to Natural Disaster* (Earthscan). Contact pelling@liv.ac.uk

**Julian E. Salt** is Senior Research Fellow, Benfield Greig Hazard Research Centre, University College London and project manager, Natural Perils at the LPC Centre for Risk Sciences. He has tracked climate change policy issues for ten years, attending the UNFCCC meetings since 1992. He specializes in the impacts of climate change on the financial sector. He is a member of the UNEP-FI Climate Change Working Group. Contact SaltJ@bre.co.uk

**Jeroen Warner** has an M.Sc. in International Relations from the University of Amsterdam. He currently co-ordinates a three-year international research project on integrated watershed management at Wageningen University, funded by the Dutch Ministry of Agriculture. He has researched and published widely on water conflict, water management and participation, and is completing his Ph.D. on images of risk and security in flood management at the Flood Hazard Research Centre, Middlesex University, UK. Contact JeroenWarner@users.TCT.WAU.NL

**Ben Wisner** is a researcher in development studies at the London School of Economics, and also has research affiliations with the Benfield Greig Hazard Research Centre and Environmental Studies at Oberlin College, his home base. He has studied disasters and hazards for 35 years, is co-author of *At Risk* (Routledge [1994] 2003), and consulted for UNDP, WHO, FAO, ILO. Contact bwisner@igc.org

# Preface

Humanitarian disasters triggered by natural hazards appear to be growing in severity and frequency of impact. For some, this is explained by natural cycles in atmospheric and geophysical phenomena. Increasingly, though this explanation is proving insufficient, not least as hydro-meteorological cycles are themselves influenced by the externalities of human development through anthropogenic global environmental change. Rather, a consensus has formed around the recognition that disaster events are simultaneously the products of inappropriate development policy, and vice versa. Dramatic examples of natural hazard events that have been allowed to lead on to humanitarian disasters because of inappropriate development include the Marmara earthquake in Turkey (1999), Hurricane Mitch in Central America (1998) and the Gujarat earthquake in India (2001). In these cases losses are high but spatially contained. However, there is a real chance that the economic losses caused by natural disasters might be sufficiently large in the near future that they could impact on regional, if not the global economy. From 1991 to 2000 natural catastrophes are reported to have caused losses of US$78.7 billion per annum, with disaster losses increasing fourteenfold between the 1950s and 1990s (IFRC/RC 2001).

The good news is that human security to so-called natural disasters is a rising political priority; the bad, that increased funds for disaster relief and engineering-based mitigation measures tend to mean a reduced budget for national and international development agencies. Juggling the balance between direct investment in disaster mitigation and response and longer-term development initiatives is tricky for politicians and aid agencies that tend to work to five-year budgeting timetables at most. In this environment there is a need for clear policy scrutiny based on a critical understanding of the relationship between natural disaster and development in a globalizing world. An initial response by theorists and practitioners to this challenge has been to revisit the Marxist interpretation of disaster first put forward in the 1980s (Hewitt 1983). However, the twenty-first century differs from the mid-late twentieth century in important ways: populations have grown, urban areas are burgeoning, capital is increasingly footloose, traditional cultures are increasingly modernized, the state is smaller and civil society actors are playing a greater role in disaster management, and we are unquestionably experiencing anthropocentric global environmental change. These changing conditions require refinements to established ways of understanding and practice in disaster and development. Marxist approaches have recently been joined by work drawing on gender studies, cultural theory, political ecology, stakeholder analysis, regime theory and structuration theory to help explain when, where and who disaster strikes.

Therefore, not only has our present era of globalization changed the pathways through which natural hazard becomes humanitarian disaster, but it has also stimulated a growing diversity of approaches to understanding the relationship between disaster and development. *Natural Disaster and Development in a Globalizing World* is designed to bring together both of these agendas of change.

*Natural Disaster and Development in a Globalizing World* does this by drawing together the work of a group of internationally recognized experts in the field of disasters and development. They include academics and practitioners with experience in economically strong (e.g. US and UK) and less economically strong (e.g. Philippines, Ecuador, Egypt, China and Turkey) countries. The book has three main aims. First, to present new perspectives on the theory and practice of disasters and development in an accessible format. The historical development of perspectives on natural disasters and human vulnerability is reviewed in the introduction in Part I. This provides a framework for the subsequent chapters that outline new divergences from or critical omissions in established theory and praxis. The second aim of the book is to present trends in globalization that require adaptation or new ways of thinking about natural disaster and disaster–development linkages. Chapters in Parts II, III and IV identify key themes linking the process of globalization with development and disaster risk at different scales: taking a global, international and local perspective in turn. This is of course no more than a presentational device and authors frequently demonstrate how processes operating at different scales interact in the production of hazard and risk. Third, the book aims to look ahead and sketch out priorities for natural disaster planning in the future. Each chapter will draw out lessons for future management and for moving the theoretical debate on disaster and development forward. A short concluding chapter compares and contrasts the conclusions of the individual chapters and offers a synthesis of these ideas. It identifies areas for future research and suggests ways in which disasters planning could adapt to respond to risk under changing global pressures.

The book is aimed at both novice (general reader, junior to mid-level students) and more expert (mid to late-level undergraduates, postgraduates and professionals) audiences. The book provides easy entry to the subject for the novice, and for the more experienced scholar/practitioner an outline of the research frontier in disaster studies. For those new to the subject, the review of disaster studies in Part I, the summary of findings in Part V and the use of dedicated bibliographies in each chapter will be of practical use. In addition to seeking out innovative contributions to disaster studies, in selecting contributors the editor was mindful of two priorities. First a wish to bring together writing from contributors based in the global North and global South. In part, *Natural Disasters and Development in a Globalizing World* represents an international dialogue on natural disasters, including contributions from countries on which little material is easily accessible (e.g. China). Second, to integrate within a single text chapters presenting views grounded in theory with those offering more empirical detail.

Mark Pelling
Liverpool

## References

Hewitt, K. (ed.) (1983) *Interpretations of Calamity from the Viewpoint of Human Ecology*, London: Allen and Unwin.

International Federation of the Red Cross and Red Crescent (IFRC/RC) (2001) *World Disaster Report 2001*, Geneva: IFRC/RC.

# Acknowledgements

This book is the product of many people's labours. Its genesis was in a joint meeting of the Mountain Research Group and Developing Areas Research Group of the Royal Geographical Society–Institute of British Geographers (RGS–IBG), held as part of the Annual Conference of the RGS–IBG and hosted by the University of Plymouth in January 2001. Thanks go to all those who helped in its organization, who contributed papers, comments or an open mind, and who braved transport disruptions and unseasonal weather(!) to make the sessions at Plymouth such a success. The collection has been inspired by many events and personalities, not least the editor's conversations with colleagues and postgraduate students associated with the MA in Globalization and Development at the University of Liverpool; with members of the RADIX website and e-list; and with the organizers and participants at two workshops on disaster vulnerability hosted by the University of Middlesex in 1999 and 2002, whose insightful comments and inspiring ideas will be seen throughout this work.

During the writing, editing and production of the book debts are owed for the support, patience and inspiration offered by Ulli Huber, for the copy-editing skills of Alan Fidler, for the organizational and editorial support offered by Jo Jacomb and Ann Michael at Routledge.

Many publishers and authors have kindly given consent for material to be reproduced in this volume and their support is gratefully acknowledged. Every effort was made to gain permissions for reproducing material; apologies for any oversight in this regard.

# Part I

# Introduction

# 1 Paradigms of risk

*Mark Pelling*

## Introduction

> Court disaster long enough, and it will accept your proposal.

Mason Cooley (1927, in Andrews *et al.* 1996), made this statement referring to social risk, but Cooley's observation can be applied equally well to highlight the self-destructive tendencies of recent past and contemporary human development. A cursory review of the rising numbers of people living in absolute poverty, of growing gaps between the rich and poor, of ongoing environmental degradation, globalizing consumer culture and of the global environmental impacts of industrialization, all set against the seeming lack of political and popular will for change, certainly makes it look like humanity has been courting disaster. That the average number of natural disasters reported world-wide per annum has doubled every decade since the 1960s (Pelling 2002a), suggests our proposition has been warmly accepted.

This book hopes to open up academic and policy discussions that link together local disasters and development with the processes and pressures of global change. This responds to rising economic and human losses to natural disaster, and the uncertainty of future risk scenarios under globalization (IFRC/RC 2001; IPCC 2001). The global scale of contemporary risk processes challenge disaster management in two ways. First, the motors of global change (e.g. past industrialization in North America and Europe) are often distanced in time and space from its impacts (e.g. contemporary sea-level rise and flooding in Bangladesh). Second, mitigating disaster requires co-ordination at the local, national and global scales. In this introductory chapter those elements of global change of most salience to the disaster–development relationship are identified, and two complementary schools of disaster analysis that have begun to examine the disaster–development relationship under global change are reviewed. Finally an outline for the book is presented.

## Natural disaster, development and global change

Work seeking to elaborate theory and develop good practice for disaster mitigation needs to take account of the interaction of natural hazards with global level change (O'Brien and Leichenko 2000). Where should we turn to base exploratory work that aims at a critical examination of the interactions between natural disaster and development in a globalizing world? There is a useful background literature on the interaction of natural disasters and development (Hewitt 1983; Varley 1994; Lewis 1999), summarized in this volume by Maureen Fordham (Chapter 4). Also of relevance is a larger literature on links between international development and global-scale processes and change (for reviews see Waters 1994; Held and Goldblatt 1999). But it is the link between natural disasters and global change, and what this might mean for human development at a variety of scales, that has received less attention and that this book seeks in a small way to examine.

### *Natural disaster*

Natural disaster is used here as shorthand for humanitarian disaster with a natural trigger. Whilst physical phenomena are necessary for the production of natural hazard, their translation into risk and potential for disaster is contingent upon human exposure and a lack of capacity to cope with the negative impacts that exposure might bring to individuals or human systems (Pelling 2001). There is no lower limit of loss for events that can be classified as catastrophic. Rather, reflecting the lack of any internationally agreed definition, and considering the scale effect of disaster, where considerable local loss may be insignificant at a larger scale of analysis, events are taken at their face value as reported in the academic literature, the media or by respondents.

### *Development*

A recent definition of development presents it as 'an economic, social and political process, which results in a cumulative rise in the perceived standard of living for an increasing proportion of a population' (Hodder 2000: 3). But this supposes a benign physical environment, allowing a cumulative rise in the standards of living. Hewitt's argument that '[i]f there could be such a thing as sustainable development, disasters would represent a major threat to it, or a sign of its failure' (1995: 155) highlights the ability of natural disasters to set back development. In the aftermath of Hurricane Mitch in Honduras and Nicaragua, human losses and damage to physical infrastructure, housing and crops have set back development efforts by a decade or more (Bradshaw *et al.* 2002). Similarly, cycles of growth and recession in the global economy destabilize linear economic growth models (Wallerstein 1980). Those excluded from cycles of cumulative development include the 84 per cent of the global population living in Africa, Asia and Latin America and the substantial minority populations of the richer northern countries that have been unplugged from global circuits of accumulation and appear left behind by progress.

Disquiet about the nonlinear relationship between modern development and risk – where despite economic growth, greater levels of development can mean higher exposure to environmental hazard and potential or actual reversals in the quality of life – has generated vibrant discussion amongst sociologists, geographers and ecological economists (Beck 1992; Peet and Watts 1996; Bryant and Bailey 1997; Castree and Braun 2001). This debate hangs on the extent to which the dominant forms of modern development are to blame for increasing global inequalities in access to basic needs, in raising human exposure to environmental risk and on their social and geographical distribution. On the one hand, theorists have tried to outline ways in which economic practices contribute to climate change; for example, Mason (1997) and Robbins (1996) examined the contributions of transnational corporations. On the other hand, work has tried to identify the potential impact of climate change for individual economic sectors, particularly agriculture (Fischer *et al.* 1994; Reilly *et al.* 1994). In disasters studies, work in the neo-Marxist tradition by, for example, Susman *et al.* (1983) and Hewitt (1983) advanced a similar agenda, arguing that material security and prosperity for some was bought at the systemic production of greater risk exposure for others. This argument linked the political-economy roots of underdevelopment with determinants of exposure to natural disaster risk. For example, international disaster aid was critiqued for reproducing pre-disaster conditions of underdevelopment and so contributing to cycles of disaster and underdevelopment. O'Brien and Leichenko (2000) reach a similar conclusion from an assessment of the likely future impacts of global climate change and economic globalization. They suggest that many regions will experience 'double exposure', being both most at risk from future disaster and also unlikely to benefit from economic globalization. In this volume, Wisner (Chapter 3) extends a neo-Marxist analysis to the diverse ways that globalized forces can shape disaster risk.

### *Globalization*

Most aspects of globalization are disputed (Giddens 1998). At a basic economic level a truly global economy has not yet emerged, with most trade continuing at the regional level, for example within the European Union (EU), and trade barriers continuing (notably excluding the importation of goods from so-called developing countries into the EU and USA), with the effect that free trade is only partial (Hirst and Thompson 1996). Politically, despite the rise of supranational actors the nation-state remains at the centre of formal development planning; but globalization in terms of increasingly global flows of people, goods and information is unquestioningly underway (Castells 1996). Growth in rates of foreign direct investment and the fluidity of money transfer brought about by electronic communications technology and labour migration are exemplars. These processes have implications for local and global development patterns and the geography of risk and disaster. Dore and Etkin (Chapter 5) examine the influence of these and other variables on the potential for countries to adapt to future natural hazard.

It should be remembered that globalization processes are nothing new for many communities and nations. Wallerstein (1980) demonstrated the world-wide reach of European economies dating to the fifteenth and sixteenth centuries. Frank (1971) showed the inequalities inherent in such global relationships, with dominant nation-states holding the power to shape unequal trade relations and forge economic dependency amongst undeveloped economies. Such patterns of dependency have contributed to risk from natural disaster in the past and today. Pelling and Uitto (2001) point to high historical losses from isolated and ill-prepared populations exposed to hurricanes in the Lesser Antilles, but suggest that contemporary vulnerability, stemming from relative economic poverty linked to histories of colonialism and continuing inequalities in the terms of global trade, continues to exist. However, there are key differences between colonial and imperial world economies of the past and present-day globalization.

The most clearly visible difference is the impact of transport, communication and information technologies on reducing the effect of space and time as constraints on human transactions and decision-making. Giddens marked this as the defining feature of modern globalization (1990). The geographer David Harvey (1989) described this as time–space compression, a reduction in the experiential distance between different points in space. The impact of time–space compression is profound for natural disaster risk. Not only does it contribute indirectly through the (re)production of geographies of economic and social power, and hence human vulnerability across the globe and locally, but it also impacts directly through the ways in which natural disaster is perceived and responded to. Indirectly, globalization is linked to the production of vulnerability – for example, by the operation of the World Trade Organization (WTO) and its relationship with transnational corporations (TNCs). Chomsky (1999) argued that TNCs have their rights protected by the WTO, which is empowered to take action against states that infringe the tenets of free trade (as understood by the WTO!). In practice, Chomsky submits, this can act not only to undermine national sovereignty and the democratic process but also tends to suppress labour rights, women and children's work codes and land reform, all of which have been shown to increase local vulnerability. Set against this negative process supporters of globalization argue that foreign direct investment and corporate social responsibility can offer a mechanism and scope for adaptation to avoid natural hazard and for enhancing local socio-economic development (DFID 2000). But this claim is not borne out by the evidence; although foreign direct investment rose from US$25 billion in 1990 to US$110 billion in 1996 just ten countries have accounted for three-quarters of all private capital flows to the developing world since 1990 (French 1998). Where international private sector operations are present they seldom become involved in activities that reduce vulnerability amongst their employees or society at large (Twigg 2001).

Beck (1992) presents an account of globalization using risk as the central organizing metaphor. He argues that in the modernization phase of development people had been prepared to ignore or accept social and ecological side effects of development in return for an improved material quality of life, but that in the present period of late modernity this preference has been undone. Beck argues that for affluent societies of the so-called developed world, risk rather than basic need fulfilment is now the central concern of individuals and policy-makers. There are a number of limitations to Beck's Risk Society thesis (Lash *et al.* 1996), not least that it cannot easily be extended to cover the majority of the world's population (in so-called developed as well as developing countries) for whom basic need fulfilment remains a priority in addition to their desire to escape environmental risk. This said, Beck does uncover the knee-jerk reflexivity of the development process to disaster with policy-makers preferring emergency response to long-term vulnerability and risk reduction measures. Beck argues that the global nature of risk requires concerted international action. Kelman (Chapter 7 in this volume) takes this argument a step further by examining the influence of natural disaster on international diplomacy and the potential for disaster events to act as bridges for building trust between previously antagonistic states.

In the social science literature globalization is often reduced to its economic, cultural or political elements and interpreted as the global spread of market liberalism and Western-style democracy. In contrast, physical science discusses the nature of global environmental change. Here (drawing on Pelling and Uitto 2001) a simple framing devise is proposed that unpacks and integrates both the human and physical components of global change. This frame incorporates observations by Kasperson *et al.* (1995) and Blaikie *et al.* (1994), who separately identified eight key global forces in the production of vulnerability: global patterns of information flow, access to the world market, urbanization, population growth, global economic pressures, environmental degradation, global environmental change and war.

Table 1.1 does not attempt to present a definitive list of global pressures, and particular elements could arguably fall into more than one category. In each disaster event, specific global pressures will come to the fore. The aim is rather to offer a flexible device to structure discussions of the relative importance of elements and their interaction with development and disaster. There is insufficient space here to review each element in turn (see Pelling and Uitto 2001), but some highlighting is worth while. Those pressures closest to the global idea include global climate change and associated sea-level rise, and international regulatory institutions such as the WTO, World Bank, International Monetary Fund and the United Nations system and global political agreements, including a range of UN sponsored conventions, the most fundamental of which is the Universal Declaration of Human Rights. Processes which are theoretically global in scope but which tend to

*Table 1.1* Components of global change

| *Global processes* | *International linkages* | *Local events occurring world-wide* |
|---|---|---|
| Global climate change<br>Sea-level rise<br>International regulatory institutions and agreements | International migration<br>Development aid flows<br>Foreign direct investment<br>Debt repayment<br>International communications<br>Cultural interaction<br>International policy co-operation | Urbanization<br>Environmental degradation<br>Identity politics<br>Nationalism<br>Insurance flight |

Source: Based on Pelling and Uitto (2001).

be experienced at the level of interaction between two or more nation-states have been categorized as international linkages. Following Castells's (1996) metaphor, many of these linkages can be thought of as cross-border flows of one kind or another – of migrants, development aid, foreign direct investment, debt repayment, international communications and cultural transfers. These aspects of globalization have the potential to tie states closer together, but they also bring greater complexity and diversity to social systems. Transnational populations, living but not fully integrating into host countries whilst retaining ties with home countries, and the cultural hybridity they play a part in creating, display such complexity. They have resilience from access to international social networks, but vulnerability comes from potential exclusion from community-level disaster preparedness or response. The third category of global pressure includes those events that individually are local in character but which in aggregate have a global impact on social and physical systems. Urbanization, the rise of nationalistic politics, insurance flight and environmental degradation such as deforestation, fresh water use and soil depletion are examples of this. In these cases potential exists for global responses to enhance local coping capacity. The UN Agenda 21 framework for action, although not legally binding, is an example of a global response to local events (Pelling 2002b). In this collection, Neil Adger and Nick Brooks (Chapter 2) unpack the relationship between global climate change and local experiences of disaster, and Julian Salt in Chapter 8 examines the strategy taken by the global insurance industry in the face of mounting financial losses to natural disaster events world-wide.

## New directions in disaster and development

Recently, reflecting diversification in social science research methodologies in general, a growing number of new approaches have been applied to studies of disasters and development. In this section a brief outline of the dominant schools of thought from the 1980s to the 2000s is provided and new innovations in disasters studies are introduced from work conducted within the remits of political ecology and complexity theory.

Changes in the theoretical approaches to disaster and development reflect the broader dynamics of social science and world politics. The 1980s and 1990s were dominated by two schools of thought: the neo-Marxists and the Behaviouralists. The neo-Marxist tradition was critical of the apparently non-political orientation of the Behaviouralists, who concentrated their efforts on understanding the ways in which individuals and groups responded to disaster events (Quarantelli 1978). This downplayed the role of social structures in shaping vulnerability. In taking their approach, the Behaviouralists saw people as influencing disaster in only a limited way, with policy recommendations focusing on disaster response and recovery. The neo-Marxist perspective (Hewitt 1983) was to view disasters as deeply embedded within the social structures that they believed shaped everyday development experience. This view allowed policy to incorporate preparedness and to engage with the ongoing struggles for resources and political influence that shaped environmental risk. However, neo-Marxism was criticized for over-privileging economic class in its analysis and for failing to identify the importance of individual agency in the production of vulnerability.

A way beyond this impasse was first hinted at by Sen (1981) in his economic analysis of famines and fleshed out more fully by Drèze and Sen's (1989) identification of the importance of individual entitlements to resources based on access to economic, political and social power in shaping vulnerability to hunger and famine. But it was not until the work of Blaikie *et al.* (1994) that the entitlements approach was widely introduced to a range of natural hazard types. Blaikie *et al.* (1994) also made the contribution of bringing social science analysis closer to physical science inputs, partly echoing the human ecology school with a firm belief in the interdependence of human and physical systems. Beginning with the entitlements perspective, the turn of the millennium has seen a reinvigoration of theoretical approaches to the study of disasters and development. This has followed from new developments in feminism, management science, international relations and institutional economics, and been stimulated by a resurgence in the recognition within human geography and development studies of the criticality of disaster–development within a globalizing world. The following sections use the broad categories of political ecology and complexity theory to structure an account of recent innovations in disaster–development theory.

### *Political ecology*

In the contemporary world of social science intense interest in the search for universal laws and global processes has often reduced the level of interest in the specific and regional. Yet it is in the complex interlocking of variables within particular places with their own unique combinations

> of environmental and human systems that the real world of environmental change is being played out.
>
> (Bradnock and Saunders 2000: 85)

Bradnock and Saunders derive this conclusion from a study of environmental change in Bangladesh. In applying a political ecology approach, Bradnock and Saunders (2000) successfully show the depth of uncertainty that remains within environmental and physical science looking for methods to predict future hazard risk, and the space this creates for ideology to impose itself on development and disaster mitigation policy and practice. Stepping back from assumptions about the neutrality of scientific method and knowledge or received wisdom surrounding environment/society interaction is essential if disaster studies is to contribute to the prediction and explanation of natural hazard risk. Recent work within a political ecology framework has approached this challenge from two complementary directions: work that examines the political context of human interaction with the environment and work that deconstructs and challenges the dominant discourses that frame prevailing understandings of environmental risk.

### *Political context*

Examining the political contexts in which certain environmental crises have been situated has drawn predominantly from cases of land management and rural social contexts in Latin America, Africa and Asia or from analysis of global level challenges (see Blaikie and Brookfield 1987; Dauvergne 1998). Exceptions include Swyngedouw's (1997) work on the political ecology of urban water systems in Guayaquil, which demonstrated the close ties between local political elites, the global economy and urban access to drinking water; and Pelling's (1999, 2002a) work on the political ecology of flood hazard in contrasting urban systems. Pelling focused on the notion of adaptive potential; the capacity of an individual, community or larger social group to respond successfully to meet a hazard threat. In this volume Allen (Chapter 11) examines the micro-politics surrounding community involvement in disaster management in the Philippines, and Özerdem (Chapter 13) studies the development failures that led to the Marmara earthquake in Turkey.

### *Disaster discourse*

The querying and attempted reshaping of accepted environmental narratives by political ecologists has revealed the political orientation of what have too often been presented as the ideologically neutral scientific or managerial components of development and environmental management policy. In particular, work has shown how dominant global discourses have tended to be projected from the global North onto the global South (Stott and

Sullivan 2000). Leach and Mearns's (1996) study of land management and soil loss in sub-Saharan Africa is an excellent exemplar questioning the received wisdom that soil loss is an outcome of overly intensive grazing. Discourse analysis is the favoured tool of this branch of political ecology and it has been ably deployed to unpack narratives surrounding sustainable development (Dryzek 1997) and global environmental change (Adger *et al.* 2000). Less common are attempts to unpack narratives lying at the centre of natural hazards debates. Movements in this direction have been made. Both Bankoff (2001) and Heijmans (2001) have examined the ways in which different political actors have influenced the meanings ascribed to the term 'vulnerability'. Bankoff argued that 'tropicality, development and vulnerability form part of one and the same essentialising and generalising cultural discourse that denigrates large regions of the world as disease-ridden, poverty-stricken and disaster-prone' (Bankoff 2001: 19). This dominant perspective paints the inhabitants of such regions as being incapable of removing themselves from danger and privileges Western expertise as the magic that can provide security in place of misery. Hewitt (1997) raises a similar criticism of the global disaster mitigation industry, calling for greater and more effective participation from those most at risk. In this volume Morris (Chapter 10) and Homan (Chapter 9) examine contrasting social constructions for disaster events in Ecuador, Egypt and England.

### *Diversity and complexity*

The attractiveness of complexity theory lies in its offering of alternative to linear explanations of cause and effect that are being undone by the unpredictability of human and physical systems tied to global processes of change. Complexity provides a meta-framework for pulling together elements from the diverse approaches to disaster that have been outlined above. Under complexity, adaptation is brought to the fore. Adaptation has been defined by complexity theorists as: 'any open-ended process by which a structure evolves through interaction with its environment to deliver a better performance' (Coveney and Highfield 1996: 118). For societies at risk from natural disaster (and other social, economic and political forms of shock and stress) the ability to adapt is critical. In this volume, Warner (Chapter 12) investigates the constraints on systems adaptation leading to flood hazard in the Netherlands and Bangladesh. The argument of complexity theory is that a component of a complex system can adapt because of the ability to self-organize, to respond to external stimuli without external coercion. Over time the number of components exhibiting successful adaptations grows so that there is potential for the whole system to change in adaptation. The ability to adapt is built on the capacity of a system's components to acquire and use resources, including energy and information, on the degree of interaction between components and on the kinds of rules that govern interaction and resource access.

It is argued that there are no simple cause and effect relationships between external stimuli and adaptations in component behaviour, with any component of the system having the potential to respond to a given change in the environment in a number of different ways, including stasis. In complexity jargon, these are traits of nonlinearity. Feedback is a key motor for nonlinearity and is the pathway through which components, and finally entire systems, are seen to exhibit potentially extreme sensitivity to variations in initial conditions (Rihani and Geyer 2001). Feedback surfaces elsewhere in risk society theory in which a distinction is made between reflexivity (knee-jerk reactions) and reflection (thought-through responses), with the former being predominant in the context of human adaptation to environmental risk (Beck 1992).

To function efficiently a system needs to find a balance between self-organization and regulation, the so-called 'edge of chaos'. For this to be achieved a system's components require information on the changing external environment and internal conditions. This in turn is more likely in systems that enable continuous learning rather than control to support collective action. Complexity acknowledges that in human systems decision-making is context-dependent, so that decisions will be constrained not only by a component's information-collecting and handling capacity but also by a range of other competing priorities. For the system as a whole, the likelihood of co-ordinated or at least complementary adaptation is enhanced when a system's population has a common knowledge base and shares social values whilst still displaying sufficient diversity to enable innovative adaptations to common stimuli. Rihani (1999) argues that many of the characteristics of so-called developing countries move them away from the edge of chaos. State repression of whole populations, women, children, ethnic or religious minorities restricts freedom to interact, learn and adapt. Illiteracy, illness, lack of access to basic needs and eroded stocks of social capital similarly act to constrain people's capability to act.

### *Complex systems and disaster management*

Comfort (1999), uses complexity theory to frame an assessment of systems response to earthquake shocks; information flow and use is found to be critical. She proposes that its use could be enhanced in two ways: first, by the application of information and communication technologies; second, by the inclusion of local knowledge in disaster response actions. Implicit in this is an understanding that disaster response is the business of local as well as national and international actors (the system components), and that it cuts across public, private and civil sector divides. The timely exchange of information between actors is facilitated by social and human capital, as well as information infrastructure, in what is termed a sociotechnical system. It is the capacity to exchange and act on incoming information across different levels that determines the self-organizational capability of a disaster response system.

Actors are more likely to respond collectively to disaster shocks when there is a pre-existing level of mutuality or social cohesion and inter-organizational support, elsewhere termed 'cognitive and institutional social capital' (Pelling 2002a), within the system. Complex systems are unable to undo historical system change, and this too can be seen as an advantage when past disaster experience has led to creative adaptation that eases the way for future adaptations. As Comfort (1999: 21) notes: 'the efficacy of that [disaster response] system however, depends very much on the initial conditions of knowledge, training, communication and economic resources available in the community prior to the event'. When collective learning and adaptation in the sociotechnical system occurs it is argued that the locus of authority will be seen to shift from centralized hierarchical structures to flatter, more participatory structures, that the focus of disaster mitigation practice will move from measuring outcomes to the monitoring of processes and that human cognitive capacity will be extended through the application of information technology.

There is no space here to go into the detail of Comfort's methodology or findings, but this work is useful in highlighting the positive and negative potential that complexity analysis can offer disasters studies. Amongst the advantages of a complexity approach is its high level of abstraction, which allows comparison of disaster systems operating within different political regime forms and as part of contrasting histories of development. In addition to Comfort's work, research into local/national scale adaptation to global climate change has used many of the elements of complexity theory without directly naming it as an influence. Lorenzoni *et al.* (2000) use the language of coevolution (Norgaard 1994) to mean a system in which human institutions and physical components interact and influence each other through feedback loops over time. Cash and Moser (2000) focus on scale as a barrier to effective communication between scientists and policy-makers at scales from the local to the global and argue that this is a constraint on self-adaptation.

There are many echoes within complexity of innovations that might otherwise remain theoretically unconnected. The central importance of information and social connections can draw fruitfully on new work exploring institutional dynamics (see Chapters 11 and 12, this volume), social capital (Pelling 1998) and disaster discourse (Chapter 9, this volume). But there are difficulties and dangers in this approach. Complexity acknowledges that systems are not closed, but the data needed to map fully the evolution of such systems with interactions and flows of energy, resources and information taking place at scales from the interpersonal to the international is very demanding. Perhaps more worryingly, because of the importance of information for complexity theory, there is a tendency in complexity work to prioritize information technology and posit policy recommendations of a technical rather than political nature, when both are necessary for long-term adaptation. Indeed there is an implicit assumption that information and communications technology are non-political elements of a disaster response system. This is unlikely; for individual organizations the introduction of such technology will likely bring about changes in internal structure and culture, whilst for the system technological enhancement can do little if political will is lacking.

## Organization of this book

The body of the book is divided into five sections. This introductory chapter forms Part I. In Part II, contributors discuss the ways in which changing global dynamics might influence geographies of natural disaster risk and capacity for human adaptation. In Part III, a number of linkages that connect global scale flows and processes with local conditions are examined. In Part IV, local context is explored through six case studies of disaster–development, each study using a contrasting research framework and methodology; a secondary aim of this section is to present innovative approaches to disaster studies Finally, in Part V, a conclusion summarizes the findings of the volume and offers some thoughts on future research directions.

## References

Adger, W.N., Benjaminsen, T.A., Brown, K. and Svarstad, H. (2000) *Advancing a Political Ecology of Global Environmental Discourses*, CSERGE Working Paper GEC 2000–10, University of East Anglia.

Andrews, R., Biggs, M. and Seidel, M. (1996) *The Columbia World of Quotations*, New York: Columbia University Press.

Bankoff, G. (2001) 'Rendering the world unsafe: "vulnerability" as Western discourse', *Disasters* 25 (1), 19–35.

Beck, U. (1992) *Risk Society: Towards a New Moderntity*, London: Sage.

Blaikie, P. and Brookfield, H.C. (eds) (1987) *Land Degradation and Society*, London: Methuen.

Blaikie, P., Cannon, T., Davis, I. and Wisner, B. (1994) *At Risk: Natural Hazards, People's Vulnerability, and Disasters*, London: Routledge.

Bradnock, R.W. and Saunders, P.L. (2000) 'Sea-level rise, subsidence and submergence: the political ecology of environmental change in the Bengal delta', in P. Stott and S. Sullivan (eds) *Political Ecology: Science, Myth and Power*, London: Arnold.

Bradshaw, S., Linneker, B. and Zúniga, R. (2002) 'Social roles and spatial relations of NGOs and civil society: participation and effectiveness post-hurricane "Mitch"', in C. McIlwaine and K. Willis (eds) *Challenges and Change in Middle America*, Harlow: Prentice-Hall.

Bryant, R.L. and Bailey, S. (1997) *Third World Political Ecology*, London: Routledge.

Cash, D.W. and Moser, S.C. (2000) 'Linking global and local scales: designing dynamic assessment and management processes', *Global Environmental Change* 10, 109–120.

Castells, M. (1996) *The Rise of Network Society*, Oxford: Blackwell.

Castree, N. and Braun, B. (2001) *Social Nature: Theory, Practice and Politics*, Oxford: Blackwell.

Chomsky, N. (1999) *Profit Over People: Neoliberalism and Global Order*, London: Turnaround Publishers.

Comfort, L. (1999) *Shared Risk: Complex Systems in Seismic Response*, New York: Pergamon Press.

Coveney, P. and Highfield, R. (1996) *Frontiers of Complexity*, London: Faber and Faber.

Dauvergne, P. (1998) 'Globalisation and deforestation in the Asia–Pacific', *Environmental Politics* 7 (4), 114–135.

Department for International Development (DFID) (2000) *Eliminating World Poverty: Making Globalisation Work for the Poor*, London: DFID.

Drèze, J. and Sen, A. (1989) *Hunger and Public Action*, Oxford: Oxford University Press.

Dryzek, J.S. (1997) *The Politics of the Earth: Environmental Discourses*, Oxford: Oxford University Press.

Fischer, G., Frohberg, K., Parry, M.L. and Rosenzweig, C. (1994) 'Climate change and world food supply, demand and trade: who benefits, who loses?', *Global Environmental Change* 4 (1), 7–23.

Frank, A. (1971) *Capitalism and Underdevelopment in Latin America*, Harmondsworth: Penguin.

French, H. (1998) 'Foreign investment in the Developing World: the environmental consequences', in L. Brown and E. Ayres (eds) *The World Watch Reader on Global Environmental Issues*, London: Norton.

Giddens, A. (1990) *The Consequences of Modernity*, Cambridge: Polity Press.

Giddens, A. (1998) *The Third Way: The Renewal of Social Democracy*, Cambridge: Polity Press.

Harvey, D. (1989) *The Condition of Postmodernity*, Oxford: Blackwell.

Heijmans, A. (2001) *Vulnerability: A Matter of Perception*, Benfield Greig Hazard Research Centre, University College London, <http://www.bghrc.com> (accessed 17 June 2002).

Held, D. and Goldblatt, D. (1999) *Global Transformations: Politics, Economics and Culture*, Oxford: Polity Press.

Hewitt, K. (ed.) (1983) *Interpretations of Calamity from the Viewpoint of Human Ecology*, London: Allen and Unwin.

Hewitt, K. (1995) 'Sustainable disasters? Perspectives and powers in the discourse of calamity', in J. Crush (ed.) *Power of Development*, London: Routledge.

Hewitt, K. (1997) *Regions of Risk: A Geographical Introduction to Disasters*, Edinburgh: Longman.

Hirst, P. and Thompson, G. (1996) *Globalization in Question*, Cambridge: Polity Press.

Hodder, R. (2000) *Development Geography*, London: Routledge.

Intergovernmental Panel on Climate Change (IPCC) (2001) *Climate Change 2001: Impacts, Adaptation and Vulnerability*, <www.ipcc.ch/> (accessed 17 June 2002).

International Federation of the Red Cross and Red Crescent (IFRC/RC) (2001) *World Disasters Report 2001*, Geneva: IFRC/RC.

Kasperson, J.X., Kasperson, R.E. and Turner, B.L. III (eds) (1995) *Regions at Risk: Comparisons of Threatened Environments*, Tokyo: United Nations University Press.

Lash, S., Szerszynski, B. and Wynne, B. (1996) *Risk, Environment and Modernity: Towards a New Modernity*, London: Sage.

Leach, M. and Mearns, R. (1996) *The Lie of the Land*, Oxford: James Currey.

Lewis, J. (1999) *Development in Disaster-Prone Places*, London: IT Publications.

Lorenzoni, I., Jordan, A., Hulme, M., Turner, R.K. and O'Riordan, T. (2000) 'A co-evolutionary approach to climate change impact assessment: Part 1. Integrating socio-economic and climate change scenarios', *Global Environmental Change* 10, 57–68.

Mason, M. (1997) 'A look behind trend data in industrialization', *Global Environmental Change* 7, 113–127.

Norgaard, R.B. (1994) *Development Betrayed: The End of Progress and a Coevolutionary Revisioning of the Future*, London: Routledge.

O'Brien, K.L. and Leichenko, R.M. (2000) 'Double exposure: assessing the impacts of climate change within the context of economic globalization', *Global Environmental Change* 10, 221–232.

Peet, R. and Watts, M. (1996) *Liberation Ecologies: Environment, Development, Social Movements*, London: Routledge.

Pelling, M. (1998) 'Participation, social capital and vulnerability to urban flooding in Guyana', *Journal of International Development* 10, 469–486.

Pelling, M. (1999) 'The political ecology of flood hazard in urban Guyana', *Geoforum* 30, 249–261.

Pelling, M. (2001) 'Natural disasters?', in N. Castree and B. Braun (eds) *Social Nature: Theory, Practice and Politics*, Oxford: Blackwell.

Pelling, M. (2002a) *The Vulnerability of Cities: Natural Hazard and Social Resilience*, London: Earthscan.

Pelling, M. (2002b) 'The Rio Earth Summit', in V. Desai and R.B. Potter (eds) *The Companion to Development Studies*, London: Arnold.

Pelling, M. and Uitto, J. (2001) 'Small island developing states: natural disaster vulnerability and global change', *Environmental Hazards* 3, 49–62.

Quarantelli, E.L. (ed.) (1978) *Disasters: Theory and Research*, London: Sage.

Reilly, J., Hohmann, N. and Kane, S. (1994) 'Climate change and agricultural trade: who benefits, who loses?', *Global Environmental Change* 4, 24–36.

Rihani, S. (1999) 'Complexity in the development process', Unpublished Ph.D. thesis, University of Liverpool.

Rihani, S. and Geyer, R. (2001) 'Complexity: an appropriate framework for development?', *Progress in Development Studies* 1 (3), 237–245.

Robbins, P. (1996) 'TNCs and global environmental change', *Global Environmental Change* 6, 235–244.

Sen, A. (1981) *Poverty and Famines*, Oxford: Oxford University Press.

Stott, P. and Sullivan, S. (eds) (2000) *Political Ecology: Science, Myth and Power*, London: Arnold.

Susman, P., O'Keefe, P. and Wisner, B. (1983) 'Global disasters: a radical interpretation', in K. Hewitt (ed.) *Interpretations of Calamity: Viewpoints from the Perspective of Human Ecology*, London: Allen and Unwin.

Swyngedouw, E.A. (1997) 'Power, nature and the city: the conquest of water and the political ecology of urbanisation in Guayaquil, Ecuador', *Environment and Planning A* 29, 311–332.

Twigg, J. (2001) *Corporate Social Responsibility and Disaster Reduction: A Global Overview*, Benfield Greig Hazard Research Centre, University College London.

Varley, A. (ed.) (1994) *Disasters, Development and Environment*, New York: John Wiley.

Wallerstein, I. (1980) *The Modern World-System II*, New York: Academic Press.

Waters, M. (1994) *Modern Sociological Theory*, London: Sage.

# Part II

# Global processes and environmental risk

# 2 Does global environmental change cause vulnerability to disaster?

*W. Neil Adger and Nick Brooks*

## Contested global environmental change

We have experienced natural hazards since the beginning of history. To those who have experienced them the world has appeared to cave in when a natural catastrophe occurs. It is only with the advent of economic, social and environmental globalization that we have, in effect, created the ability to actually make our world cave in and to change it irrevocably. At the same time, globalization makes us more aware of the impacts of natural hazards, and our perceptions of risk from them cannot be divorced from its social setting. The physical basis of many natural hazards is assumed to be periodic but essentially in equilibrium. Some elements of the natural world that wreak havoc when they occur do so with unpredictable timing, but are in themselves predictable after a fashion and at other timescales. But some hazards are changing in nature due to global environmental change. In this chapter we seek to elucidate what global environmental change might mean in the context of globalization, to outline some examples of global environmental change and the implications for exposure to natural hazards, and to examine some evidence of whether there has been a change in the scale and scope of environmentally 'triggered' natural hazards in the past century.

The term 'global environmental change' is contested and problematic. First, all forms of environmental change are in some sense global or, more accurately, universal. This issue is not merely semantic, but rather frames the way in which risk and response to environmental change are perceived, particularly at the level of public policy. Second, and related, the term 'global environmental change' has become synonymous with a mindset that sees the transnational nature, or global public-good nature, of environmental change as the justification for exclusively global and market-oriented solutions to solve them. This is common across the range of so-called global environmental problems, from biodiversity loss to desertification and climate change (Adger *et al.* 2001).

In the 1990s the discourses of global environmental change have moved to the centre ground of environmental debates, leading to global-scale solutions for what are perceived to be significant environmental problems. At

*Table 2.1* Types of global environmental change

| *Type* | *Characteristic* | *Example* |
|---|---|---|
| Systemic | Direct impact on globally functioning system | (a) Industrial and land-use emissions of greenhouse gases<br>(b) Industrial and consumer emissions of ozone-depleting gases<br>(c) Land-cover changes in albedo |
| Cumulative | Impact through world-wide distribution of change | (a) Groundwater pollution and depletion<br>(b) Species depletion/genetic alteration (biodiversity) |
| | Impact through magnitude of change (share of global resources) | (a) Deforestation<br>(b) Industrial toxic pollutants<br>(c) Soil depletion on prime agricultural lands |

Source: Kasperson *et al.* 2001: 3.

the heart of this realization of the global significance of environmental change are the two major environmental problems of depletion of the stratospheric ozone layer and the issue of global climate change. The nature of these two phenomena spurred the framing of international environmental agreements, scientific networking and political consensus-building within the scientific community (see Clark *et al.* 2001 for example) on a scale unprecedented before then. These two issues, perhaps climate change in particular, have had remarkable political and economic ramifications. But added to these came a myriad of so-called global environmental phenomena for which global science and global policy-making, and the implications for environmental risk and hazard, are hugely different. In Table 2.1 a range of global environmental problems are classified in a two-way typology of systemic and cumulative change, proposed by Turner *et al.* (1990) and discussed and updated by Kasperson *et al.* (2001). Systemic risks are those which impact on an environmental system operating at the planetary scale; cumulative global environmental change is that which becomes important because it occurs everywhere.

Environmental threats come in many guises. Although much work on the consequences of global change for natural hazards has been undertaken in the context of global climate change, it is clear that the state of vulnerability to external stress is often determined by the coincidence of external environmental threats with social and economic changes in the political economy context. Local extinction of key species associated with natural ecosystem change, health impacts of changing patterns of disease, degradation of groundwater resources, forest cover change and a variety of other environmental changes in many ways eclipse systemic global change, at least in the short run, in terms of their contribution to the onset of natural disasters.

These other environmental threats are more localized but no less complex in terms of causes, thresholds and dynamic impacts. Kasperson and colleagues

(2001) argue that both systemic and cumulative environmental changes pose significant challenges in the understanding of both the drivers and processes of change, but also in the predictability of change and the risks involved, given a trajectory of change. At the same time, however, it needs to be recognized that global environmental change is not a given environmental phenomenon – in the case of climate change, upward trends in emissions of greenhouse gases and inexorable rises in temperature and sea level are not predetermined. These are socially determined futures over which there is a degree of space for action. Hence social vulnerability to these global environmental risks is a construct of the physical and the social worlds.

The second problem with global environmental change is the underpinning of discussion and action in the field by an unshakeable belief in the necessity of global-scale action to the exclusion of locally determined sustainable development priorities. It has been argued that this is, in part, a result of scientific advance in detecting global environmental change phenomena such as climate change, and the rise of global-scale scientific endeavours (Shackley and Wynne 1995). Other motivations may simply be associated with economic and cultural globalization. Goldman (1998) and others question this hegemony of global environmental change. Goldman argues that the science contributing to this debate is simply another facet of a new dominant paradigm where commons management is a panacea and antidote to all threats. Further, he argues that the emerging global commons paradigm is driven by interests that seek to colonize and extract from global commons that were previously only locally controlled.

In summary, the term 'global environmental change' is intuitively appealing as it captures some depressing trends in the state of the world's environment. Some of these phenomena may be more accurately described as universal rather than global. We further recognize that the term has particular purchase in some areas of public policy where perceptions of global action have become institutionalized. We believe that long-term environmental change, particularly climate change, can add stress to resource and human systems and can in addition become a major driving force in the changing nature, intensity and frequency of natural disasters.

## Environmental triggers for natural disasters?

### *Causality and change*

A society experiences a natural disaster when it is subject to an environmental perturbation of such a magnitude that its ability to cope is exceeded. Internal changes within a society may be such that this coping ability is reduced, rendering it more vulnerable to events that are relatively common. Such changes may be social, economic or institutional in nature, or may be the result of a society's interaction with its local physical environment. For example, the impact of extreme rainfall events may be heightened by deforestation that increases runoff and leads to flooding. Alternatively, an individual event may be of such a magnitude that it overwhelms an otherwise functioning and healthy society that has not experienced a reduction in its coping capacity. By definition, such events are unusual, either representing long return-period extremes of natural variability, or resulting from environmental change. Examples of long return-period extremes are particularly severe storms, earthquakes and volcanic eruptions, which may recur on multi-century or millennial timescales. Many examples of such discrete events may be found in twentieth-century records, although such events do not fall into any of the categories in Table 2.1 as they are not the result of continuous changes in the ambient physical environment but, rather, are manifestations of natural variability.

Examples of natural disasters arising from continuous large-scale environmental change are most likely to be found in the study of climate change. Humans have adapted to long-term changes in climate and other environmental parameters. Indeed, the history of agriculture can be seen in terms of adaptation of cropping and livestock systems to an ever-changing set of circumstances (e.g. Bray 1986; Diamond 1999). Nevertheless there are examples where climate change may take place so rapidly that societies are unable to adapt to it. Such disasters may be viewed as examples of systemic changes as described in Table 2.1. With hindsight it is possible to identify past natural disasters as consequences of global climate change. For example, the desiccation of the present arid belt stretching from West Africa to China, some 5,000 years ago, was associated with large-scale societal disruption and the rise and fall of civilizations as people either adapted or failed to adapt to arid conditions (Roberts 1998). While climatic determinism has remained unpopular in many quarters, largely due to its association with nineteenth- and early twentieth-century pseudo-science concerning differences between races (see, for example, Huntingdon 1924), there is increasing evidence that large-scale, systematic changes in global climate have had profoundly negative consequences for many societies in the past (Keys 1999; Cullen *et al.* 2000; de Menocal 2001). We can expect current and future systemic change to have similarly dramatic consequences for societies throughout the world, mediated by a process of economic globalization that may magnify or mitigate disruption for any given population group depending on its relationship with the rest of the globalized world. However, without the benefit of hindsight it is extremely difficult to attribute any particular climatic event to systematic climate change.

Climate and weather variability are significant constraining factors in human interventions with the biotic world for agricultural purposes. Agriculture is a major sector through which climate plays a role in economic development, and there is a long history of analysis of the adaptation of human societies to different climates and to changes in climates in terms of food production (Lamb 1995, for example). At the global scale, a further analogy of global warming as a coevolutionary process has been set out by Schneider and Londer (1984). Human societies have always altered their own environments. The scale of alterations in recent times, through land use change in the past few centuries (Richards 1990) and through industrialization since the nineteenth century, has altered the composition of the atmosphere and hence the global climate (Houghton and Skole 1990). No part of the world is immune to these changes in the global climate system (the climate is a pure public good in the economic sense). Many parts of the world are affected by feedbacks in the climate system, to which human society is forced to adapt. Adaptation may further alter the ambient environment in order to maintain 'equilibrium' in resource use. This global coevolution of the climate system with the economic system therefore involves inanimate parts of nature, not just the biotic world.

The most important aspects of the causal relationships between climate and social systems are the scale of analysis and the directness or remoteness of the links between climate and human activities. Thus, for example, social upheaval and ultimately peasant rebellion against colonial French rule during early twentieth-century Vietnam occurred concurrently with severe floods and droughts throughout the region (Scott 1976; Adger 1999). The link between typhoon flooding and agricultural output is clear, while the impact of reduced harvest with high taxes as a trigger for rebellion is obviously more speculative. Ingram *et al.* (1981) point out that regarding scale, the inevitable concentration on acute periods of crisis may be misleading. Short-term influences of climate, at the annual or intra-annual scale, do not necessarily imply any long-term causal relationship between climate and human history. Thus the key issues in analysing the role of climate in shaping human society and the evolution of economies and institutions are the attribution of causality and definition of appropriate scale in the analysis. Technologies, land use and institutions themselves evolve. The driving forces of this evolution are demand for the goods and services provided by land; induced innovation of technologies; and social, political and cultural factors (see Norgaard 1994). Allied to these, climate makes some contribution to, or acts as a constraint to, particular activities and technologies in many resource-dependent societies.

Cumulative environmental change as described in Table 2.1 may be the direct result of human manipulation of the local surface environment or of climate change, a systemic process which may in turn be caused by anthropogenic modification of the global atmospheric environment. Cumulative and systemic changes may interact to exacerbate environmental problems, as in the deforestation/flooding example given on p. 21. Deforestation may also interact with drought to cause soil degradation and desertification, and the removal or die-back of vegetation due to human activity or climate change may produce new sources of carbon, thus further exacerbating global climate change (Williams and Balling 1996). Norberg-Bohm *et al.* (2001) summarize the causal structure of environmental hazards in terms of changes in material fluxes caused by natural processes or human activity, which in turn cause changes in *valued environmental components* (VECs) or ecosystem services.

Changes in direct and indirect services modulate the exposure of human populations and the systems on which they depend, and therefore have consequences for these populations and systems. Changes in material fluxes may involve processes such as the redistribution of carbon between the ocean or land surface and the atmosphere, the release of pollutants into groundwater, the removal of soil via wind or water erosion, or changes in plant and animal stocks. Ecosystem services are those systems that humans value for practical or aesthetic purposes, and exposure represents the pathways via which changes in functions and services impact on human health, economic well-being, physical infrastructure and ecosystems.

Systemic and cumulative environmental change are intimately linked via complex feedback processes, the most obvious of which are land–atmosphere interactions that involve changes in quantities such as heat and moisture fluxes, rainfall, vegetation cover, soil erosion, atmospheric circulation and global atmospheric greenhouse gas and aerosol content. Rainfall, vegetation, soils and local and regional climate regimes are all examples of VECs, and changes in them affect valued systems by altering exposure to droughts, floods, famines (via reduced agricultural productivity) and other disasters.

Below we explore the relationship between systemic and cumulative environmental change in the African Sahel, and examine their impacts on human populations within the context of vulnerability to famine. The Sahel is chosen as it has experienced marked changes in rainfall amounts and variability over the course of the twentieth century, and consequently has been widely studied. Changes in rainfall have been associated with widespread suffering related to drought and famine, which has been exacerbated by local and global socio-economic trends. However, in recent years there is evidence of effective adaptation to a changed environment.

### *Environmental change in the Sahel*

The Sahel, situated at the southern fringe of the Sahara and stretching from the West African coast to the East African highlands (Figure 2.1), is particularly prone to drought. The region achieved international prominence during the catastrophic drought of the early 1970s during which hundreds of thousands of people and millions of animals died (de Waal 1997; Mortimore 1998). Dry conditions began in the late 1960s and have persisted until the present day, with some amelioration during the 1990s (Figure 2.2).

The increased frequency of dry years after the late 1960s represents a significant desiccation of the Sahelian region (Hulme 1996). This desiccation has been associated with a shift from interannual rainfall variability to quasi-decadal scale variability, and an increase in rainfall persistence, or the degree to which one year's rainfall resembles that of the previous year (Brooks 2000). An important control on rainfall in the Sahel appears to be the contrast in temperature between the northern and southern hemispheres as reflected in patterns of sea-surface temperature (SST) anomalies: when the northern hemisphere oceans are cold relative to the southern hemisphere

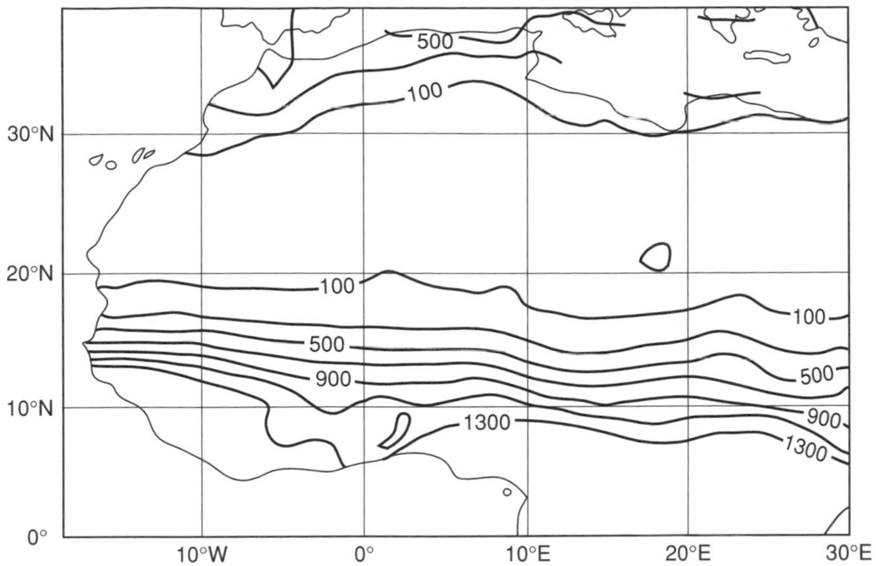


*Figure 2.1* Isohyets representing mean annual rainfall in mm over northern Africa for the period 1901–1996. The Sahel corresponds approximately to the zone of high south–north rainfall gradients where annual rainfall amounts average 100–700 mm. The data were obtained from the Climatic Research Unit and are described in New *et al.* (1999).

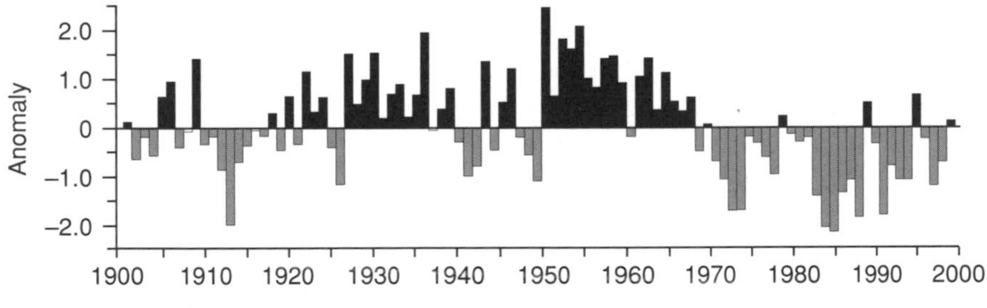


*Figure 2.2* Spatially aggregated annual rainfall anomalies (in standard deviations) representing the region 10–20°N; 25°W–30°E, roughly corresponding to the Sahelian zone. Anomalies are calculated with respect to the mean for the entire series (1901–1998).

oceans and northern Indian Ocean drought is more common in the Sahel (Folland *et al.* 1986; Street-Perrott and Perrott 1990; Ward *et al.* 1993). While the world as a whole has warmed over the twentieth century (IPCC 2001a), from the late 1950s until the mid-1970s northern hemisphere SSTs cooled. Southern hemisphere SST anomalies (relative to the 1961–1990 average) exceeded those of the northern hemisphere from the early 1970s until around 1990 (IPCC 2001a). This pattern may have been caused by a number of factors. While such fluctuations may be manifestations of internal oceanic or atmospheric variability, they may also be caused by the following phenomena: (1) reduced solar heating of the northern hemisphere oceans due to atmospheric pollution caused by human activity and increased dust emission from deserts, particularly the Sahara (Charlson *et al.* 1992; N'Tchayi *et al.* 1994; Schollaert and Merrill 1998); (2) a reduction in the strength of the thermohaline circulation (THC) that drives the Gulf Stream, bringing warm tropical surface waters to the North Atlantic (Hansen *et al.* 2001); or (3) injections of cold fresh water resulting from the melting of sea ice or permafrost, which may in turn also suppress the deep-water formation that drives the THC (Hecht 1997; Cavalieri *et al.* 1997). It should be noted that the northern Indian Ocean, cut off from northern high latitude influences, behaves in the same way as the southern hemisphere oceans. Warming of the northern Indian Ocean intensifies local large-scale convection, and may consequently intensify large-scale subsidence over West Africa via the Walker circulation, suppressing convection and therefore rainfall over the Sahel (Shinoda and Kawamura 1994). Dry years in the Sahel are also associated with a weakening of the south–north circulation over northern Africa, which has been synchronous with tropospheric warming (Shinoda 1990).

All of the above are observed climatic changes, and all are consistent with scenarios of human-induced climate change as described in modelling studies (e.g. IPCC 1996). It is therefore highly plausible that they represent manifestations of systemic change driven by emissions of greenhouse gases from twentieth-century industrialization. Global economic activity may well have contributed significantly to the recent environmental changes that have occurred in the Sahel, and which have been associated with widespread human suffering and societal disruption via drought and famine.

It is interesting to note that, despite decades of controversy over locally driven environmental change in the Sahel, there is little or no evidence that cumulative change in the form of overgrazing, deforestation or other 'inappropriate land use practices' has played a role in bringing drought to the Sahel. However, prolonged drought episodes may increase the likelihood that resource exploitation becomes unsustainable, increasing vulnerability, as the land surface is denied the opportunity to recover from dry periods in inter-drought years. Dry conditions appear to be predominantly driven by exogenous processes, and even changes in dust-storm activity, usually explained in terms of land degradation and desertification, may be at least partly due to changes in atmospheric circulation (Brooks 2000; Brooks and Legrand 2000).

### *Environmental change as a factor in the 1972/3 Sahelian famine*

The Sahel drought of 1972/3 was the culmination of a downward trend in rainfall that commenced in the 1950s, when rainfall throughout the region was high. By 1972, a large number of Sahelians had already suffered from several years of drought. The only previous year during the period of meteorological records in which aggregated rainfall for the whole Sahel region was similarly deficient was 1913 (Figure 2.2). The cumulative effect of drought in the late 1960s and early 1970s was to increase people's vulnerability in the short term by depleting their stocks of capital, grain and animals, and undermining their health as well as reducing the amount of available rural labour by encouraging migration (Rau 1991; Cross and Barker 1992).

Although the famine of the early 1970s was precipitated by several consecutive years of drought, a number of non-climatic factors had conspired to make societies particularly vulnerable to drought. As in most famines, it was the rural poor who suffered most. The following factors all contributed to the vulnerability of rural communities through political marginalization and decreased food security: (1) isolation due to poor communications and transport links, (2) an urban bias in policy-making resulting from poor rural representation, (3) a focus on short-term economic stabilization rather than long-term development, (4) an emphasis on industrial investment and the conversion of agriculture to cash crops at the expense of the production of food for local consumption (Baker 1987; Shaw 1987; Rau 1991). Lack of democratic decision-making structures and the prioritization of economic growth through industrialization and agricultural exports, partly as a result of pressure from foreign creditors, contributed to the above (Rau 1991; Sen 1999).

Sahelian governments also encouraged nomadic populations to settle in highly marginal areas that, although cultivable in the wet decade of the 1950s, quickly reverted to aridity in the subsequent drought. Populations in such regions were nomadic in order to cope with marginal conditions and highly variable rainfall. Where they were not settled, nomads were pushed into even more marginal areas by the expansion of agriculture, reducing their coping options in times of drought (Horowitz and Little 1987).

It appears that a combination of sustained drought, altered demographics, social engineering and external political and economic factors had increased vulnerability to drought in the Sahel during the early 1970s.

### *Adaptation*

The Sahel has not experienced a repeat of the massive, systematic, regional-scale famine that occurred in the 1970s, despite continued rainfall scarcity (Mortimore 2000), although food scarcity has continued to affect populations in some locations (Cross and Barker 1992). Regional rainfall deficits were greater than or comparable to those of the early 1970s in 1983, 1984, 1987 and 1990 (Figure 2.2), and some locations continued to experience severe rainfall deficits throughout the 1990s (Mortimore 2000). Evidently it was not drought alone that led to the widespread human suffering of the early 1970s: the socio-economic factors that led from drought to famine via increased vulnerability have been briefly summarized above.

As well as disrupting societies on a large scale, drought and famine can also act as a trigger for adaptation. Because of its location in the transition zone between humid equatorial Africa and the hyper-arid Sahara, the Sahel is highly sensitive to variations in global and regional climate which modulate the West African Monsoon. As a consequence of the resulting high climatic variability, its inhabitants have been developing highly effective strategies to cope with drought since the region became semi-arid some four or five thousands of years ago (Andah 1993; Casey 1998). These strategies, including agricultural diversification, labour export and migration (e.g. Rain 1999) have survived despite centuries of foreign interference and decades of state neglect and/or mismanagement (e.g. de Waal 1997). While the factors listed earlier in this chapter mitigated against successful short-term adaptation to persistent drought conditions in the early 1970s, the massive societal disruption caused by the famine paradoxically appears to have both stimulated and facilitated local autonomous adaptation. The changed socio-economic landscape of the region provided opportunities, and change was also encouraged by other external factors.

For example, the removal of state subsidies on artificial fertilizers encouraged people to turn to integrated crop and livestock management, increasing animal numbers in order to provide manure, which is used instead of artificial fertilizer (Mortimore 2001). In many locations cash crops have been replaced by food crops, and more resilient crop varieties have been introduced (Mortimore and Tiffen 2001). Soil conservation and well-managed tree plantations are also emphasized (Mortimore 2001). These local-level adaptation strategies have enabled farmers to increase soil fertility and maintain or even increase food production despite reduced rainfall and a shortening of the growing season (Rain 1999; Mortimore 1998, 2001). Indeed, even for cash crops, which are traditionally viewed as environmentally damaging, evidence from Mali and elsewhere suggests that cotton and other cropping systems are adaptable. Cash crop farmers act to minimize livelihood risks from their exposure to economic as well as environmental shocks (Benjaminsen 2001).

Adaptation has not been restricted to farm management. Income diversification has also played a part in reducing vulnerability to drought (Mortimore 2001). Although the growth of cities is associated with a host of problems, urban centres provide a market for locally produced foodstuffs and surplus farm labour. Migration and labour export have always been coping strategies in the Sahel, with migration peaking in drought years (Rain 1999). However, they are not simply measures of last resort, but represent an important source of income for rural communities wishing to diversify their resources and reduce their vulnerability to drought (Mortimore and Tiffen 2001).

### *Globalization and vulnerability: lessons from the Sahel*

In this chapter we explore the contribution of environmental change to vulnerability to natural disasters. The Sahelian famine of the early 1970s certainly qualifies as a natural disaster, given that it was precipitated by drought and killed large numbers of people and animals. Furthermore, the drought was a manifestation of regional and possibly global climate change. The increased frequency of dry years after the onset of desiccation and the shift to quasi-decadal rainfall variability undoubtedly increased people's vulnerability in the short term, as repeated consecutive dry years meant that there was little opportunity to recover from one drought and prepare for the next.

Referring to rainfall changes and agriculture in the Kano-Maradi region which straddles the borders of Nigeria and Niger, Mortimore (2000) states that 'Climate change has not undermined the basis of economic life. Positive trends in soil fertility and tree management, in a context of intensifying drought stress, are remarkable in such circumstances.' Whereas unusually persistent drought may increase people's vulnerability in the short term, in the medium to long term it may encourage adaptation. Rather than causing vulnerability, environmental change in the Sahel has increased the *exposure* of the population to one particular type of natural disaster: drought, and, by extension, drought-associated famine. Vulnerability is not simply a function of exposure but also of people's capacity to adapt to change. If the latter remains unchanged, increased exposure will lead to increased vulnerability.

The lessons of the Sahel are common throughout history and throughout the world. Vulnerability is caused by inequality, inappropriate governance structures, and maladaptive economic and agricultural development. In the Sahelian case, an emphasis on urban industrialization in countries with predominantly rural economies, and an emphasis on cash crops and export earnings in countries where much of the population cannot afford or access imported foodstuffs, is a recipe for increased national vulnerability. Paradoxically, it may require the shock of a catastrophic natural disaster, or a shift away from policies originally designed to encourage national economic growth, before national populations can begin to adapt by developing a sustainable rural economy. Most often, though, this opportunity comes from political reform rather than from natural disaster. The rebuilding of Honduras after Hurricane Mitch is a prime example where an apparent 'clean slate' does not necessarily lead to reform – the status quo in uneven development is simply replicated (Glantz and Jamieson 2000).

The Sahelian example also demonstrates the importance of local-level adaptation, and illustrates how people may adapt effectively outside of a global commons management framework. The maladaptive nature of policies ostensibly designed to increase national prosperity demonstrates the danger inherent in a 'top-down' approach to development, particularly when it is dictated by the global economic paradigms of the day, paradigms that often exist in a vacuum as far as considerations of the physical world and rural communities are concerned. But all countries are subject to the vagaries of globalization. Market liberalization and decentralization of governance structures in Vietnam during the 1990s have exacerbated vulnerability to typhoon hazard in that country because the basis for co-ordinated local level collective action has been undermined (Adger 2000).

The determination of developing countries' national economic and social policies by remote institutions driven by a particular ideology, such as the current free-market ideology promulgated by the rich nations and the global economic institutions, therefore holds dangers for developing countries with large populations of rural poor who are largely excluded from the global economic system. However, globalization may also provide opportunities for income diversification and access to markets, decreasing vulnerability to famine. There is evidence that urban markets created by oil revenues, and derivative incomes, acted to mitigate vulnerability to famine in northern Nigeria in the 1970s (Mortimore 1998; but see Watts 1983). These findings suggest that globalization may benefit the poor of the developing world in some particular instances as long as it allows them sufficient space for autonomous adaptation, and does not encourage states to undertake maladaptive social and economic engineering projects. Nonetheless, globalization in any form is likely to produce winners and losers as long is it is associated with changes in existing orders.

## Trends in twentieth-century climate-related disaters

It is in the analysis of trends in natural disasters and their economic and human costs that the consequences of combinations of external environmental and internal socio-economic factors become most apparent. For example, data from the insurance industry indicate that global economic losses from catastrophic weather events increased by a factor of ten between the 1950s and the 1990s; the number of events associated with these losses increased by more than a factor of five (IPCC 2001b). However, these increases are not necessarily due to changes in climatic regimes or the frequency of extreme events, and may be explained at least partially by changes in recording practices or the extent of insurance cover.

The Emergency Events Database (EM-DAT) of international disasters provides a further means of analysing frequencies and impacts of disasters related to extremes in climate. This dataset is widely used in the field of disaster studies, and is available from the Centre for Research on the Epidemiology of Disasters (CRED) at the Université Catholique de Louvain in Brussels (http://www.cred.be/emdat). While EM-DAT nominally represents the entire twentieth century, data coverage is poor prior to 1970; data before this date should be treated with caution, a conclusion also drawn by other authors (e.g. Charvériat 2000).

Figure 2.3a–d shows global annual incidences of floods, windstorms, epidemics and landslides (including avalanches), the four disaster types that exhibit the most striking trends in the EM-DAT database over the final decades of the twentieth century. Other types of disaster exhibit considerable variability. Droughts and wildfires exhibit a peak in the early to

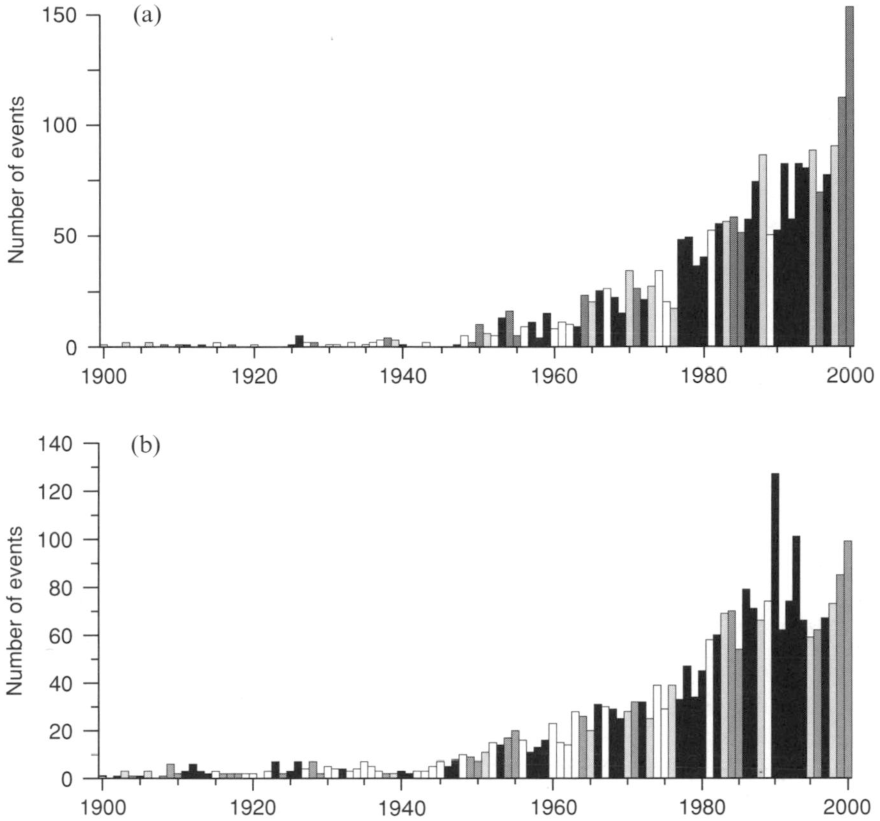


*Figure 2.3* Annual global occurrences of four disaster types. Four types of shading are used in this figure. El Niño years are represented by dark bars with lighter shading for years effected by either a transition from or into an El Niño or La Niña year, non-ENSO years are clear. Note the different scales on the vertical axes. (a) Recorded floods, (b) Recorded windstorms, (c) Recorded epidemics, (d) Recorded landslides.

mid-1980s and another around the turn of the century (Figure 2.4); a large component of the 1980s drought maximum is likely to be a result of the drought in the Sahel, which affected a number of countries. Recorded heat waves and cold waves increase in frequency after 1980, and insect infestations exhibit a large peak in the late 1980s. All categories except insect infestations exhibit a large frequency maximum for the year 2000, and of these the year 2000 is associated with the largest number of occurrences except in the case of droughts and windstorms, for which it is exceeded only by 1983 and 1990 respectively.

All disaster types, except drought and insect infestations, exhibit upward trends in frequency of occurrence. Aggregated disasters show an upward trend interrupted by a slight decline from the late 1980s until the late 1990s

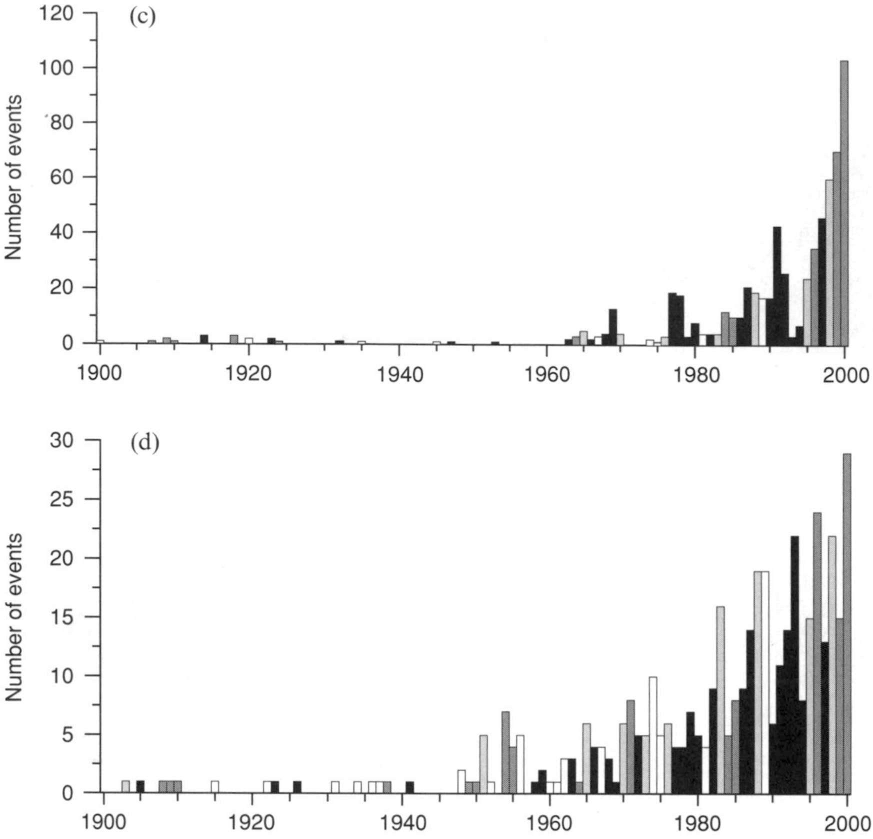


*Figure 2.3* (*cont'd*)

(Figure 2.5). This feature is encouraging, as it suggests that the frequency of recorded disasters is not merely rising in parallel with constantly improving recording practices. Similarly, the peak in drought frequency at the height of the Sahelian dry period indicates that the data do indeed reflect variability in extreme event frequency.

It is difficult to ascertain whether the above trends in recorded disaster frequencies are the result of changes in the return periods of weather extremes, better reporting practices, or other factors. An event is much more likely to be recorded if it affects large numbers of people, or the physical infrastructure or economy of a country. As populations grow and expand into areas that are more climatically marginal or more exposed to weather extremes, climatic variability is more likely adversely to affect human beings or the systems on which they depend. Numbers of people killed and otherwise affected (injured, displaced or requiring immediate assistance) by climate-related disasters exhibit a trend similar to those in Figure 2.3. However, it

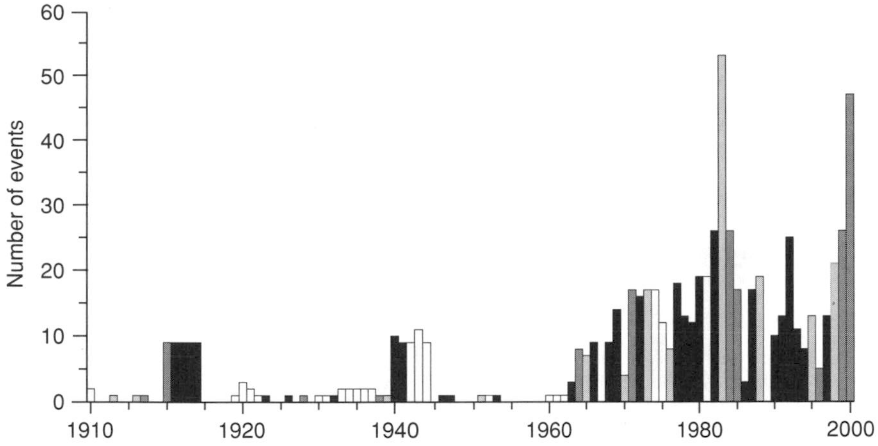


*Figure 2.4* Global drought occurrence. Note a maximum at the time of the driest period in the Sahel. Four types of shading are used in this figure. El Niño years are represented by dark bars with lighter shading for years effected by either a transition from or into an El Niño or La Niña year, non-ENSO years are clear.

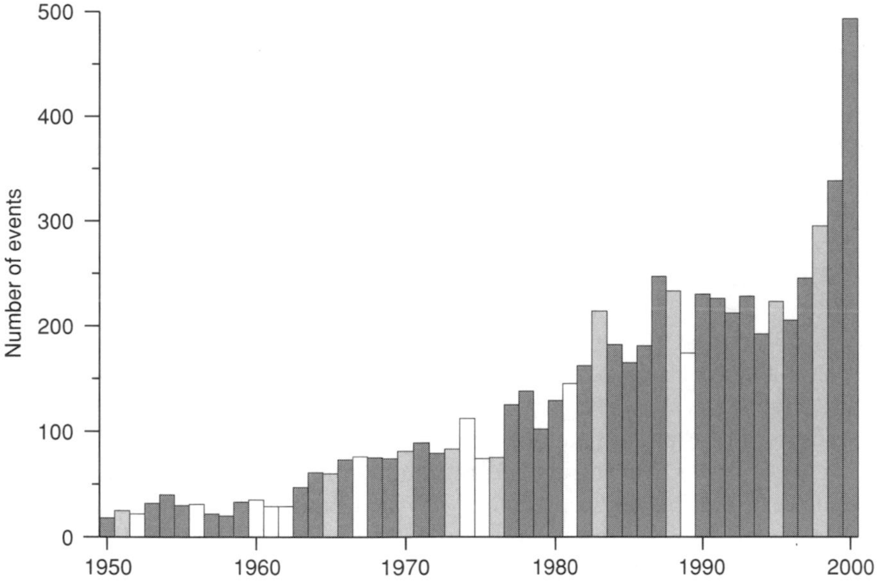


*Figure 2.5* Global frequencies of all disaster types. Four types of shading are used in this figure. El Niño years are represented by dark bars with lighter shading for years effected by either a transition from or into an El Niño or La Niña year, non-ENSO years are clear.

is interesting to note that four years are associated with greater numbers of killed and otherwise affected than the year 2000, and the largest entry is for 1987.

It is probable that the increased frequency of recorded disasters results from a combination of climatic change and socio-economic and demographic changes. While the evidence for climatically driven changes in disaster frequency is limited, the IPCC (2001a) reports changes in some extreme climate phenomena. The Third Assessment Report (TAR) states that 'it is likely that there has been a widespread increase in heavy and extreme precipitation events in regions where total precipitation has increased, e.g. the mid- and high-latitudes of the northern hemisphere' (IPCC 2001a: 163). However, it reports that there is little evidence for increases in tropical storm intensity and frequency, and uncertainty as to whether mid-latitude cyclones have exhibited any such trends (IPCC 2001a: 163). Observed increases in areas classified as severely wet and severely dry are closely related to the shift in El Niño–Southern Oscillation (ENSO) towards more warm events since the late 1970s and coincide with record high global mean temperatures (IPCC 2001a: 162). However, Figure 2.3 and other series (not shown) indicate that there is no clear systematic relationship between ENSO events and the annual global frequency of climate-related disasters.

The IPCC (2001a) report that there is no evidence for widespread systematic changes in severe local weather events such as tornadoes, thunder days, lightning or hail, although changes have occurred in some regions. Trends represented by increases in minimum temperatures, maximum temperatures, number of hot days, and heat index (a combined measure of heat and humidity related to human comfort) are reported as 'likely' or 'very likely', and an increase in continental drying and drought risk is reported as 'likely in some areas' (see Table 2.2). Records of hazards such as floods, slides, wildfires, epidemics and insect infestations are not addressed directly in the TAR, but such events are likely to be associated with increases in precipitation and continental drying as described above.

Nonetheless, the relative importance for disaster frequencies of twentieth-century climate change when compared with changes in human vulnerability is still unclear. The period of maximum global warming, i.e. since 1970 (IPCC 2001a), has also been a time of accelerating globalization during which humanity has become collectively more aware of global environmental problems and human vulnerability. The discovery of the ozone 'hole' alerted national governments and the public in the developed world to the existence of global environmental problems, and interest in climatic and environmental change has been sustained by bodies such as the Intergovernmental Panel on Climate Change (IPCC) and events such as the UN Earth Summit in 1992. Events such as the Sahelian famine of the early 1970s and the Ethiopian famine of the mid-1980s have led to a greater awareness of natural disasters and humanitarian crises. Factors as disparate as the formulation of the Kyoto Protocol on Climate Change and the attacks on the

*Table 2.2* Estimates of confidence in observed and projected change in extreme weather and climate events

| *Changes in climate phenomenon* | *Confidence in observed changes (latter half of twentieth century)* | *Confidence in projected changes (during twenty-first century)* |
|---|---|---|
| Higher maximum temperatures and more hot days over nearly all land areas | Likely | Very likely |
| Higher minimum temperatures, fewer cold days and frost days over nearly all land areas | Very likely | Very likely |
| Reduced diurnal temperature range over most land areas | Very likely | Very likely |
| Increase of heat index over land areas* | Likely over many areas | Very likely over most areas |
| More intense precipitation events | Likely over many northern hemisphere mid- to high-latitude land areas | Very likely over many areas |
| Increased summer continental drying and associated risk of drought | Likely in a few areas | Likely over most mid-latitude continental interiors |
| Increase in tropical cyclone peak wind intensities | Not observed in few analyses available | Likely over some areas |
| Increase in tropical cyclone mean and peak precipitation intensities | Insufficient data for assessment | Likely over some areas |

Source: IPCC (2001a, 2001b).

Note: * Heat index is a combination of temperature and humidity that measures effects on human comfort.

United States in September 2001 have raised levels of concern about poverty and social and political exclusion. At the same time, economic growth and global economic integration have increased the value of capital assets at risk from natural disasters, just as population growth has increased numbers of people at risk. All of these factors have led to greater efforts to monitor environmental variability and change and their impacts.

## Future directions

Recent positive trends in natural disaster frequency and associated economic damage are unlikely to reverse in the near future. The expansion of economic activity and settlement, as populations grow and economic globalization continues, will result in the exposure of greater numbers of people and greater quantities of capital assets to natural disasters. Where natural resources are not managed sustainably, cumulative environmental change in the form of processes such as soil loss and degradation, groundwater pollution, use of fossil water reserves, increased runoff potential due to deforestation, and overfishing will increase exposure of vulnerable communities to hazards such as food and water scarcity and flooding, and their vulnerability to drought, famine and disease. Climate change is very likely to exacerbate such anthropogenic cumulative change by increasing exposure to risk.

Although there is no conclusive evidence that recent disaster trends are the result of climate change, the global temperature record suggests that large-scale rapid anthropogenic warming has only commenced within the past several decades. We are therefore working with relatively short time series of data relating to an anthropogenically altered atmosphere resulting from the 'inadvertent biogeophysical experiment we are conducting with planetary climate' (Hulme 2002: 15) that might be altering exposure to natural disasters. These series may be too short for the identification of disaster trends, and they are made ambiguous by the factors discussed in the previous section. It is therefore instructive to turn to modelling studies and theoretical considerations in order to illuminate our understanding of possible future relationships between climate change and natural disasters. The most comprehensive assessment of potential future climate change is the series of reports of the Intergovernmental Panel on Climate Change, and the following discussion of projected changes in climate and their impact on exposure is based on the findings of Working Group I of the IPCC (2001a), unless otherwise indicated.

Although there is much uncertainty concerning future climate change, we can be almost certain that temperatures and sea levels will continue to rise in the near future. Higher baselines of temperature and sea level will increase exposure via higher frequencies of extreme events that are temperature and sea-level dependent (Figure 2.6). Increased temperatures over land will lead to more hot spells, which will increase heat stress on living organisms and also lead to enhanced evapotranspiration, thus increasing agricultural demand for water while reducing surface water availability, increasing the risk of drought. These factors may be offset by increases in precipitation, although increases in mean precipitation are most likely in middle and high latitudes. Increases in precipitation extremes are likely to be widespread; in the tropics and sub-tropics a higher frequency of extreme rainfall events is more likely to increase the number of floods and epidemics than mitigate drought, the incidence of which is likely to increase due to projected continental drying. Regions such as the Sahel are particularly vulnerable to increases in rainfall variability and extremes; the timing of rainfall is as important as the amount of rainfall for agriculture in the Sahel. Extreme rainfall events in semi-arid regions are also likely to lead to increased soil erosion; analysis of interannual variability in rainfall and atmospheric dust content in the Sahel indicates that episodic rainfall events play an important role in generating the erodible material that is necessary for the development of dust storms (Brooks and Legrand 2000). Atmospheric dust is a major cause of respiratory problems in regions such as the Sahel (Griffin *et al.*

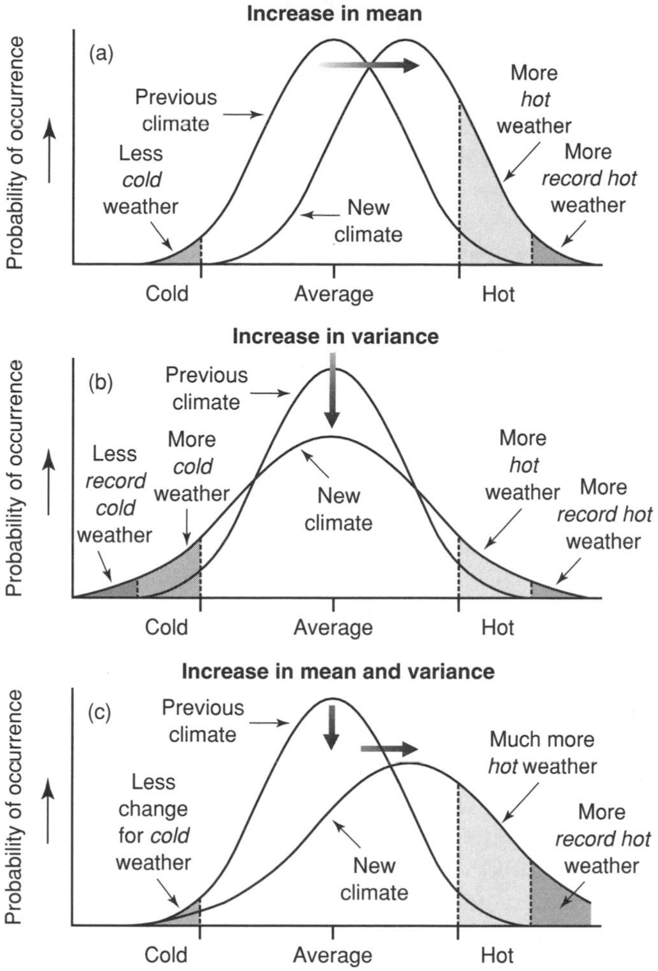


*Figure 2.6* Changes in the probability of extreme events with increases in mean and variance for temperature.

Source: IPCC (2001a).

2001), and a shift towards higher rainfall variability and intensity might therefore have a negative impact on health.

Elevated temperatures over the ocean are likely to affect the development of tropical storms. Although there is little evidence that tropical storm frequency has increased in recent decades or will increase in the near future, the IPCC (2001a) suggests that tropical storm intensities may increase, and that 'mean and peak precipitation intensities from tropical cyclones are likely to increase appreciably' (IPCC 2001: 73). This would have important implications for human mortality resulting from storm-related flooding and landslides (Cockburn *et al.* 1999). Elevated sea levels will increase the risk of flooding from storm surges for coastal communities, as well as exacerbating coastal erosion and groundwater contamination. With the number of people living below the 1,000-year flood level estimated to rise to 600 million by 2100 (Nicholls and Mimura 1998), increased exposure of coastal populations to climatic risk is likely to enhance nations' vulnerability to natural disasters by a significant amount.

The above argument concerns the consequences of rises in global mean temperature and sea level, to which we are committed because of past emissions of greenhouse gases and future emissions that will certainly continue well into the twenty-first century, increasing the greenhouse gas content of the atmosphere for years, and probably decades, to come. We can be certain that climate change will occur, and we can also be certain that population growth will increase population densities in areas exposed to climatic risk. What is highly uncertain is the magnitude, rapidity, and even nature of climate change over the coming decades and centuries. However, this uncertainty is not simply a result of our lack of understanding of the climate system, but is also a function of our lack of knowledge as to how human societies will develop. The IPCC Special Report on Emissions Scenarios (SRES) uses a number of modelling studies to demonstrate that the extent and rapidity of climate change over the twenty-first century will depend strongly on the nature of the process of globalization. Important factors include the degree and rapidity of global economic integration; the rate at which alternative energy sources are developed; the amount of emphasis on environmental protection, sustainable development and social equity; the extent to which solutions to environmental problems and poverty are local as opposed to global; and the rate of population growth. These and other factors will interact in a complex manner to determine the evolution of greenhouse gas emissions, and hence of the coupled system comprising human society, the surface environment and the global climate.

Choices made by individuals, institutions and governments over the next few decades will determine the degree and nature of future climatic and environmental change for millennia to come (Burgess *et al.* 2000). Twenty-first century choices regarding energy technologies, economic models, intellectual property and technology transfer, equity and sustainable development, and patterns of consumption will determine the exposure and vulnerability of future generations. While the process of globalization may not be inexorable, further economic integration, technological development, cultural exchange and human migration will continue in the short to medium term. The future nature of our world, globalized or otherwise, will depend on how these processes are managed. Global environmental change will be a fact of life for future generations; it will alter their exposure to environmental risk and hence partially determine their vulnerability to natural disasters. The extent of the change with which they have to cope, and the options available for minimizing their vulnerability, will be decided by the politics of today.

## References


Adger, W.N. (1999) 'Evolution of economy and environment: an application to land use in lowland Vietnam', *Ecological Economics* 31, 365–379.

Adger, W.N. (2000) 'Institutional adaptation to environmental risk under the transition in Vietnam', *Annals of the Association of American Geographers* 90, 738–758.

Adger, W.N., Benjaminsen, T.A., Brown, K. and Svarstad, H. (2001) 'Advancing a political ecology of global environmental discourses', *Development and Change* 32, 681–715.

Andah, B.W. (1993) 'Identifying early farming traditions of West Africa', in T. Shaw, P. Sinclair, B. Andah and A. Okpoko (eds) *The Archaeology of Africa: Food, Metals and Towns*, London: Routledge.

Baker, R. (1987) 'Linking and sinking: economic externalities and the persistence of destitution and famine in Africa', in M.H. Glantz (ed.) *Drought and Hunger in Africa: Denying Famine a Future*, Cambridge: Cambridge University Press.

Benjaminsen, T.A. (2001) 'The population–agriculture–environment nexus in the Malian cotton zone', *Global Environmental Change* 11, 283–295.

Bray, F. (1986) *The Rice Economies: Technology and Development in Asian Societies*, Oxford: Blackwell.

Brooks, N. (2000) 'Dust–climate interactions in the Sahel–Sahara zone of northern Africa, with particular reference to late twentieth century Sahelian drought', unpublished Ph.D. thesis, University of East Anglia. Available online at <http://www.cru.uea.ac.uk/~e118/thesis/thesis.html>.

Brooks, N. and Legrand, M. (2000) 'Dust variability over northern Africa and rainfall in the Sahel', in S. McLaren and D. Kniveton (eds) *Linking Climate Change to Land Surface Change*, Dordrecht: Kluwer Academic Publishers.

Burgess, P.E., Palutikof, J.P. and Goodess, C.M. (2000) 'Investigations into long-term future climate changes', in S.J. McLaren and D.R. Kniveton (eds) *Linking Climate Change to Land Surface Change*, Dordrecht: Kluwer Academic Publishers.

Casey, J. (1998) 'The ecology of food production in West Africa', in G. Connah (ed.) *Transformations in Africa: Essays on Africa's Later Past*, Leicester: Leicester University Press.

Cavalieri, D.J., Gloersen, P., Parkinson, C.L., Comiso, J.C. and Zwally, H.J. (1997) 'Observed hemispheric asymmetry in global sea ice changes', *Science* 278, 1104–1106.

Charlson, R.J., Schwartz, S.E., Hales, J.M., Cess, R.D., Coakley, J.A., Hansen, J.E. and Hofmann, D.J. (1992) 'Climate forcing by anthropogenic aerosols', *Science* 255, 423–430.

Charvériat, C. (2000) *Natural Disasters in Latin America and the Caribbean: An Overview of Risk*, Inter-American Development Bank Research Department, Working Paper No. 434. Available from: <http://www.iadb.org/RES/working_papers_list.cfm?CODE=WP-434>

Clark, W.C., Jäger, J., van Eijndhoven, J. and Dickson, N. (eds) (2001) *Learning to Manage Global Environmental Risks: A Comparative History of Social Responses to Climate Change, Ozone Depletion and Acid Rain*, Cambridge, Mass.: MIT Press.

Cockburn, A., St Clair, J. and Silverstein K. (1999) 'The politics of "natural" disaster: who made Mitch so bad?', *International Journal of Health Services* 29 (2), 459–462.

Cross, N. and Barker, R. (eds) (1992) *At the Desert's Edge: Oral Histories from the Sahel*, London: Panos and SOS Sahel.

Cullen, H.M., de Menocal, P.B., Hemming, S., Hemming, G., Brown, F.H., Guilderson, T. and Sirocko, F. (2000) 'Climate change and the collapse of the Akkadian empire: evidence from the deep-sea', *Geology* 28 (4), 379–382.

De Menocal, P.B. (2001) 'Cultural responses to climate change during the late Holocene', *Science* 292, 667–673.

Diamond, J. (1999) *Guns, Germs and Steel*, New York: Norton.

Folland, C.K., Palmer, T.N. and Parker, D.E. (1986) 'Sahel rainfall variability and worldwide sea temperatures, 1901–85', *Nature* 320, 602–606.

Glantz, M. and Jamieson, D. (2000) 'Societal response to Hurricane Mitch and intra- versus intergenerational equity issues: whose norms should apply?' *Risk Analysis* 20, 869–882.

Goldman, M. (1998) 'The political resurgence of the commons', in M. Goldman (ed.) *Privatising Nature: Political Struggles for the Global Commons*, New Brunswick: Rutgers University Press.

Griffin, D.W., Kellogg, C.A. and Shinn, E.A. (2001) 'Dust in the wind: long range transport of dust in the atmosphere and its implications for global public and ecosystem health', *Global Change and Human Health* 2, 20–33.

Hansen, B., Turrell, W.R. and Østerhus, S. (2001) 'Decreasing overflow from the Nordic seas into the Atlantic Ocean through the Faroe Bank channel since 1950', *Nature* 411, 927–930.

Hecht, J. (1997) 'Baked Alaska', *New Scientist* 156, 4.

Horowitz, M.M. and Little, P.D. (1987) 'African pastoralism and poverty: some implications for drought and famine', in M.H. Glantz (ed.) *Drought and Hunger in Africa: Denying Famine a Future*, Cambridge: Cambridge University Press.

Houghton, R.A. and Skole, D.L. (1990) 'Carbon', in B.L. Turner, W.C. Clark, R.W. Kates, J.F. Richards, J.T. Mathews and W.B. Meyer (eds) *The Earth as Transformed by Human Action*, Cambridge: Cambridge University Press.

Hulme, M. (1996) 'Recent climatic change in the world's drylands', *Geophysical Research Letters* 23, 61–64.

Hulme, M. (2002) 'Climate change: a sober assessment', in T. Gilland (ed.) *Hurricanes, Floods and Climate Change: Nature's Revenge?*, Abingdon: Institute of Ideas and Hodder and Stoughton.

Huntingdon, E. (1924) *Civilization and Climate* (3rd edn), New Haven: Yale University Press.

Ingram, M.J., Farmer, G. and Wigley, T.M.L. (1981) 'Past climates and their impact on man: a review', in T.M.L. Wigley, M.J. Ingram and G. Farmer (eds) *Climate and History: Studies in Past Climates and their Impact on Man*, Cambridge: Cambridge University Press.

IPCC (1996) *Climate Change 1995: The Science of Climate Change*, WMO and UNEP, Cambridge: Cambridge University Press.

IPCC (2001a) *Climate Change: The Scientific Basis*, WMO and UNEP, Cambridge: Cambridge University Press.

IPCC (2001b) *Climate Change 2001: Impacts, Adaptation and Vulnerability*, WMO and UNEP, Cambridge: Cambridge University Press.

Kasperson, R.E., Kasperson, J.X., Dow, K., Ezcurra, E., Liverman, D.M., Mitchell, J.K., Rattick, S.J., O'Riordan, T. and Timmerman, P. (2001) 'Global environmental risk and society', in J.X. Kasperson and R.E. Kasperson (eds) *Global Environmental Risk*, Tokyo: United Nations University Press and London: Earthscan.

Keys, D. (1999) *Catastrophe: An Investigation into the Origins of the Modern World*, London: Arrow Books.

Lamb, H.H. (1995) *Climate, History and the Modern World* (2nd edn), London: Routledge.

Mortimore, M. (1998) *Roots in the African Dust*, Cambridge: Cambridge University Press.

Mortimore, M. (2000) *Profile of Rainfall Change and Variability in the Kano-Maradi Region, 1960–2000*, Working Paper No. 25, Drylands Research, Crewkerne, Somerset, UK (available from http://www.drylandsresearch.org.uk).

Mortimore, M. (2001) *A Profile of Natural Resource Management in the Kano Region, 1960–2000*, Working Paper No. 37, Drylands Research, Crewkerne, Somerset, UK (available from http://www.drylandsresearch.org.uk).

Mortimore, M. and Tiffen, M. (2001) *Livelihood Transformations in Semi-arid Africa*, Working Paper No. 40, Drylands Research, Crewkerne, Somerset, UK (available from http://www.drylandsresearch.org.uk).

New, M.G., Hulme, M. and Jones, P.D. (1999) 'Representing 20th century space–time climate variability. Part II: Development of 1901–1996 monthly terrestrial climate fields', *Journal of Climate* 13, 2217–2238.

Nicholls, R.J. and Mimura, N. (1998) 'Regional issues raised by sea level rise and their policy implications', *Climate Research* 11, 5–18.

Norberg-Bohm, V., Clark, W.C., Bakshi, B., Berkenkamp, J.-A., Bishko, S.A., Koehler, M.D., Marrs, J.A., Nielsen, C.P. and Saga, A. (2001) 'International comparisons of environmental hazards', in J.X. Kasperson and R.E. Kasperson (eds) *Global Environmental Risk*, Tokyo: United Nations University Press and London: Earthscan.

Norgaard, R.B. (1994) *Development Betrayed: The End of Progress and a Coevolutionary Revisioning of the Future*, London: Routledge.

N'Tchayi, G.M., Bertrand, J., Legrand, M. and Baudet, J. (1994) 'Temporal and spatial variations of the atmospheric dust loading throughout West Africa over the last thirty years', *Annales Geophysicae* 12, 265–273.

Rain, D. (1999) *Eaters of the Dry Season*, Oxford: Westview Press.

Rau, B. (1991) *From Feast to Famine*, London: Zed Books.

Richards, J.F. (1990) 'Land transformation', in B.L. Turner, W.C. Clark, R.W. Kates, J.F. Richards, J.T. Mathews and W.B. Meyer (eds) *The Earth as Transformed by Human Action*, Cambridge: Cambridge University Press.

Roberts, N. (1998) *The Holocene: An Environmental History*, Oxford: Blackwell.

Schneider, S.H. and Londer, R. (1984) *Coevolution of Climate and Life*, San Francisco: Sierra Club Books.

Schollaert, S.E. and Merrill, J.T. (1998) 'Cooler sea surface west of the Sahara desert correlated to dust events', *Geophysical Research Letters* 25 (18), 3529–3532.

Scott, J.C. (1976) *The Moral Economy of the Peasant: Rebellion and Subsistence in Southeast Asia*, New Haven, Conn: Yale University Press.

Sen, A. (1999) *Development as Freedom*, Oxford: Oxford University Press.

Shackley, S. and Wynne, B. (1995) 'Global climate change: the mutual construction of an emergent science-policy domain', *Science and Public Policy* 22, 218–230.

Shaw, T.M. (1987) 'Towards a political economy of the African crisis', in M.H. Glantz (ed.) *Drought and Hunger in Africa: Denying Famine a Future*, Cambridge: Cambridge University Press.

Shinoda, M. (1990) 'Long-term Sahelian drought from the late 1960's to the mid-1980's', *Journal of the Meteorological Society of Japan* 68, 613–624.

Shinoda, M. and Kawamura, R. (1994) 'Tropical African rainbelt and global sea surface temperatures: interhemispheric comparison', *Proceedings of the International Conference on Monsoon Variability and Prediction,* Volume I, Geneva: WMO.

Street-Perrott, F.A. and Perrott, R.A. (1990) 'Abrupt climate fluctuations in the tropics: the influence of the Atlantic Ocean circulation', *Nature* 343, 607–612.

Turner, B.L., Kasperson, R.E., Meyer, W.B., Dow, K.M., Golding, D., Kasperson, J.X., Mitchell, R.C. and Ratick, S.J. (1990) 'Two types of global environmental change: definitional and spatial-scale issues in their human dimensions', *Global Environmental Change* 1, 14–22.

De Waal, A. (1997) *Famine Crimes: Politics and the Disaster Relief Industry in Africa*, African Rights and the International African Institute, Bloomington: Indiana University Press.

Ward, N., Folland, C.K., Maskell, K., Colman, A.W., Rowell, D.P. and Lane, K.B. (1993) 'Experimental seasonal forecasting of tropical rainfall at the UK Meteorological Office', in J. Shukla (ed.) *NATO ASI Series 16: Prediction of Interannual Climate Variations*, Berlin: Springer-Verlag.

Watts, M. (1983) *Silent Violence. Food, Famine and the Peasantry in Northern Nigeria*, Berkeley: University of California Press.

Williams, M.A.J. and Balling, R.C. (1996) *Interactions of Desertification and Climate*, WMO and UNEP, London: Arnold.

# 3 Changes in capitalism and global shifts in the distribution of hazard and vulnerability

*Ben Wisner*

## Introduction

### *'Risk society' and its others: are you feeling lucky today?*

Economic and, to some extent, political globalization have been accompanied with a lot of theory. Post-modernism can be interpreted as theorizing struggle by individuals, small self-identified groups and localities to maintain their autonomy in the face of huge, rapidly shifting tidal waves of capital. In the wake of these waves are pulled millions of international labour migrants – many of them illegal. Their enclaved communities are often abused by mainstream cultures. At best they are ignored, be they Turks or Mozambicans in Germany, Algerians in France, Salvadorans, Mexicans or Cambodians in Southern California.

The critique of modernism has attempted to understand changes in society at the macro level. Such is the work of Ulrich Beck, whose *Risk Society* (1992) remains a major reference point. He is primarily concerned with technological hazards and the perception of these by affluent Europeans. This is the domain of nuclear power, dioxin and mad cow disease. But what of the 'other' world? In February 2001 the European Union's minister of agriculture announced steps to prevent the spread of mad cow disease totalling US$1 billion a year. That same month the government of El Salvador announced that it would cost at least US$1 billion to rebuild infrastructure (roads, bridges, schools, hospitals) destroyed by the earthquakes of January and February 2001.

Beck writes about a small portion of humanity. Most of humanity is more afraid of warlords than terrorists. Many more lives are shattered by AIDS, tuberculosis and malaria than by the human form of mad cow disease. The risks of crossing the US–Mexican border, where 475 people succumbed to heat, cold, drowning, and exhaustion in 2000, are not the risks of brain cancer from cell phone use by more than 100 million nervous *Norteamericanos*. A photo in a recent newspaper showed a couple sunbathing on a beach in Spain where the body of an unsuccessful migrant had washed up.

There are truly global threats, of course, such as climate change and biodiversity erosion. In addition, there are a number of transboundary (though not fully global) technological hazards (Linnerooth-Bayer *et al.* 2000). All these, however, affect different classes and groups of people very differently. A framework is needed that can accommodate these diverse risks in an understanding of globalization and its discontents, its promises and its contradictions. This chapter makes a preliminary, tentative attempt, urged on by Hewitt's complaint (now nearly twenty years ago), that disasters are treated as though they were located outside society, on a kind of 'disaster archipelago' (Hewitt 1983).

### *Power and material interest: who's quick and who's dead?*

After the 1976 earthquake in Guatemala the word on the streets for the event was 'class-quake', because low income, indigenous people were hardest hit (Blaikie *et al.* 1994). In the aftermath of Hurricane Andrew women told researchers that they would like to invest more in making their homes safe, but their husbands would not let them (Enarson and Morrow 1997). In both these cases political economy is in evidence. What is at work is the role of differences in power and material interest in shaping the spatial and social distribution of risk. Gender relations in the household embody power and perceived differences in material interest, as suggested in the example from Hurricane Andrew. However, these gender relations were played out in a larger-scale context. This was the growth boom of south Florida in the 1980s and early 1990s, weak regulation of the building industry, downsizing and restructuring that left many working-class men anxious about future employment (Enarson and Morrow 1997; Peacock *et al.* 1997). A series of well-known international examples make clear the complex nesting of local, national and international factors.

In 1984 a Union Carbide chemical factory in Bhopal, India malfunctioned. Three thousand people died, and 30,000 were severely injured. The Bhopal tragedy came in the context of world-wide enthusiasm for hybrid seeds and the Green Revolution package that went with them. The national context was Indian modernization policy that encouraged the production of pesticides for the Green Revolution in factories like the one in Bhopal. All this was superimposed on the class and caste structure of central India, where rural displacement had led to rapid squatter settlement in cities (Shrivastava 1992; cf. Perrow 1984 and Cutter 1993).

Hurricane Mitch killed 30,000 people in Honduras and Nicaragua in 1998. This deadly result was partly the effect of displacement by agri-business of small farmers onto small subsistence farms on steep slopes. Deforestation and slope instability followed. International markets for beef, bananas, cotton and coffee drove this displacement, while historic patterns of inequality in land access and political power absorbed and translated these forces into unsafe conditions at the local level (Comfort *et al.* 1999; cf. Diaz and Pulwarty 1997). In Armenia, in 1988, 100,000 died in an earthquake. The standardized, modular, cheap apartment buildings that collapsed in Armenia were partly a result of cold war competition that diverted resources in the Soviet Union into defence spending, as well as the centralist and technocratic nature of the Soviet state. An ineffective response to the emergency was also partly shaped by centralist control of communication, as well as by a long history of cultural and linguistic mismatch between Russian-speaking officials and the Armenian-speaking population (Verluise 1989; Comfort 1999).

In none of these cases would I argue that power and material interest fully determine a disastrous outcome. Many other factors – social, technical, administrative and legal – also contributed (Wisner 1993a, 1996, 1998). However, a full understanding of such disasters is impossible without taking political economy into account.

Earthquake tragedies in El Salvador and the Indian state of Gujarat in 2001 also cry out for analysis of complex root causes. It is not part of the human condition to be buried under a landslide triggered by an earthquake. Earthquakes happen. But the disaster follows because of human action and inaction. In the case of the middle-income neighbourhood of Las Colinas in Santa Tecla, just outside the capital, San Salvador, 400 homes were lost beneath a wall of debris from a collapsing slope above. This was not an 'act of God'. A group of Las Colinas residents and environmental groups were in court in 2000 to stop development on that slope and the ridge above. The judge ruled against them. It is also not an 'act of God' that so few of the multi-storey structures in Indian cities are built according to earthquake resistant norms (Wisner 2000, 2001).

In El Salvador, and also in Gujarat, both the poor and the middle class suffered. In both places hungry rural people have been migrating in search of work to cities like San Salvador, Ahmedabad and Bhuj. They become squatters who live in makeshift dwellings in some of the most potentially dangerous areas in an earthquake. They have little or nothing to invest in making their homes safer, and little incentive to improve their dwellings because they don't own the land where they've built.

In San Salvador and Ahmedabad alike, the middle class is attracted to the rapidly growing edge of the sprawling cities (Satterthwaite 1999). Developers and contractors rush to fill this market demand, often in too much haste to observe building codes. This is where the landslide buried hundreds in Las Colinas, and where new apartment houses for Ahmedabad's salaried workers came crashing down. In El Salvador and India rural impoverishment and crisis have led to large numbers of wage migrants crowding informal settlements in the major cities. Gujarat's booming economy has attracted landless and land-poor people from all over northern and central India.

### *Disaster = failure of human development*

Academic support for the critique of a blind belief in economic growth as the sole goal for development has been building up since the United Nations Development Programme began to publish its Human Development Report

(HDR) in 1990 (UNDP 1990–2001). Its Human Development Index (HDI) measures equity, health and education, not simply economic activity. In 1995 the HDR added gender-specific measures, and in 1997 two separate measures of human poverty – one for more-developed countries and one for the less developed. Other international institutions have responded to the reintroduction of social and other human goals into the development discourse (UNRISD 2000). Finally, in 2001 the World Bank devoted two chapters to poverty and disaster vulnerability in its *World Development Report* (World Bank 2001: 135–178).

In its *World Disaster Report 2001*, the International Red Cross presented data from UNDP and the Center for the Epidemiology of Disasters (CRED) that compares the impacts of extreme natural events on countries with high, medium, and low scores on the Human Development Index (IFRC 2001: 162–165). They looked at data for 2,557 disasters triggered by natural events from 1991–2000. Half of these disasters took place in countries with a medium HDI, but two-thirds of the deaths occurred in countries with a low HDI. Only 2 per cent of the deaths were recorded in the countries with a high HDI. Tabulating deaths and monetary costs per disaster makes the relationship with HDI even clearer (Table 3.1).

UNDP took this analytical work even further in 2002 by commissioning the mathematical study of more than two hundred possible indicators of disaster risk vulnerability in coming up with an index for use in its *World Vulnerability Report*. The result was striking. The Human Development Index again turns out to be the best predictor of deaths due to extreme natural events, world-wide, on average over the twenty years 1980–1999 (UNDP forthcoming).

*Table 3.1* Level of human development and disaster impacts

| | *Deaths per disaster* | *Cost per disaster (US$m)* |
|---|---|---|
| Low HDI | 1,052 | 79 |
| Medium HDI | 145 | 209 |
| High HDI | 23 | 636 |

Source: Based on IFRC (2001: 162, 164).

## The uses of political economy

Why can an understanding of economic and political power improve routine practice by planners and NGOs? A first important reason is that implementation of ideas about disaster risk mitigation and prevention has to take place in existing economic and political circumstances. Both national and local initiatives succeed or fail to the extent that they are compatible with existing patterns of power and material interest. Consider as an example a refinery complex recently acquired in a merger by a large holding company.

This company has borrowed heavily to pay for the merger. They want to cut costs. These same executives want the restructured refinery to appear more efficient so that its stock value increases and their personal wealth in executive stock options grows. Employee downsizing is the key to cost reduction and the short-term appearance of efficiency. The refinery in-house fire brigade is eliminated. Fire fighting is now the responsibility of the local city and county. Maintenance is also cut back. Small fires are more frequent and there is the danger that they will not be contained quickly. Low income immigrant residents adjacent and down wind of the refinery are concerned but have less political voice than the absentee owners of the holding company, who make large campaign contributions to politicians.

Nothing described above is illegal, at least not in the United States. Politicians, holding company executives, refinery managers share a common belief in growth, efficiency and deregulation – key concepts of neo-liberal ideology. This is where one has to begin to use the 'C' word. One has to ask how capitalism, as a manner of organizing power and material interest, has changed over the past hundred years of so. What are the characteristics of capitalism today that contribute to an uneven spatial and social distribution of risk?

### *Neo-liberalism: on the dangers of the greed*

In the early part of the last century capitalism was characterized by a more direct relationship between production and improved quality of life. Since roughly the period when Henry Ford invented his factory system several changes have eroded the fit between capitalism's functioning and quality of life. Here are some of the principle characteristics of what Pope John Paul has called 'savage' capitalism. Others have used these adjectives: 'global', 'foot loose', 'post-Fordist', 'wild', 'rampant', 'triumphant', 'predatory' (Korten 1995; Mander and Goldsmith 1996; Corbridge *et al.* 1994; Harvey 1982, 1996, 2000, 2001; UNRISD 1995). The UN Institute for Social Development Research has called it 'globalization with a human mask' (UNRISD 2000).

First, many workers today are not able to consume the goods and services they help to produce for export or for their own country's upper middle class and elite. At the extreme, there are many rural people in the world who produce food and fibre for the world market but are unable adequately to feed themselves. The young women making Nike shoes in Vietnam will never save enough to own such footwear themselves. In the US and parts of Europe, global competition, industrial downsizing, out-sourcing, the rise of the service economy and de-unionization have combined to increase the number of temporary, part-time, and minimum wage workers (Chossudovsky 1997; Sassen 1998).

Second, the link between capitalist production and general well-being has been severed by the growth of military and luxury production. The percentage of capitalist production devoted to such 'waste' has increased since Henry Ford's day, accelerating during the Reagan/Thatcher/Kohl cold war end game. Workers can't get to work in Trident submarines or B-1 bombers, and fur coats don't trickle down (O'Connor 1994; Daly and Cobb 1989). Dreams of a 'peace dividend' have now totally evaporated in a frenzy of new military spending in the 'war on terrorism'.

Third, contemporary capitalism is committed to maximizing growth irrespective of negative long-term environmental consequences. The US has refused to ratify the Kyoto Accord on green house emissions. In the United States and much of Europe development continues to eat away at peri-urban land. Growth without regulation seems to have become a core value (Daly and Cobb 1989; Durning 1992; Jaeger 1994; O'Connor 1994; Sachs 1999). One insightful study of this process in El Salvador juxtaposes 'sterile growth' with 'development', meaning human or social development (Rubio *et al.* 1997).

The consequences for hazard mitigation of these trends are numerous. First, overseas predatory foreign and national capital rapidly deplete forest cover, drill for oil, delve for minerals, set up sweat shops and plantations for cheap exports, and the workers and rural people are faced with vast environmental destruction, insecure, low-wage jobs, and hazardous working and living conditions. It is the rare Third World state that has been able to contain the negative effects of such 'growth', despite great social unrest. An indirect, but important, consequence for the US is that a large proportion of aid money that could have gone for sustainable development overseas is absorbed by disaster foreign assistance (Bales 1999; Reed 1996; Johnston 1997, UNRISD 1995, 2000).

Second, in the US growing polarization between the rich and poor (including many immigrant workers) means that community based approaches to disaster preparedness and mitigation will find it more difficult to get representation and volunteer activity from the least affluent and marginalized end of the income distribution. Even the middle class is working harder according to Harvard economist Juliet Schor, and charities have noted a decline in volunteerism across the board (Schor 1996; Mander and Goldsmith 1996).

Third, the economically marginal are unlikely to be able to respond to incentives (e.g. lower insurance rates, subsidies) for retrofitting their residences (Mileti 1999).

Fourth, much of the growth taking place world-wide, and in the US in particular, is perverse. That is, it is growth that does not increase general welfare in the long run. Addition of the 'n-th', incremental shopping mall, multiplex cinema, golf resort, office park creates profits for a few, distraction for some, stress for many and planning nightmares for those concerned with transportation, evacuation, other infrastructure, service provision and open space. Much critical infrastructure is already under-maintained due to cut backs in public finance at many levels. Sprawl adds new infrastructure in need of future maintenance, without which it may collapse or malfunction in an extreme event (Durning 1992; Weizsaecker *et al.* 1997; Ayres and Weaver 1998; Burby 1998; Godschalk *et al.* 1999; Inoguchi *et al.* 1999; Mitchell 1999; Abramovitz 2001).

One of the most important ways economic globalization is redistributing risk is via international and internal migration. For example, in the Pico Union district of the City of Los Angeles there are several thousand indigenous Maya from Guatemala, many of whom do not speak Spanish, let alone English. They are fugitives of the genocide against rural people during the worst of Guatemala's civil war in the 1980s. They live in crowded and poorly maintained tenements and work in sweatshops that manufacture clothing. Most are undocumented immigrants. They are an invisible and relatively powerless group of human beings caught in the gears of a complex mechanism that links US historical support for Central American oligarchies and their military forces with neo-liberal regimes of accumulation. These people are highly socially vulnerable to a variety of health hazards and also earthquake and building fire (Wisner 1999).

Gujarat provides an example of how internal migration affects the social and spatial distribution of risk, and how economic development creates such migration flows. This Indian state is an economic powerhouse. As such, one sees there in microcosm the enormous gap between rich and poor that characterizes India itself and the world. Much of this economic growth is linked to global markets. Its successful economy has come at the cost of having to accommodate somewhere a very large population of unskilled labour. These people have migrated there from all over northern India because their lives as landless labourers elsewhere were untenable.

Gujarat is to be the major beneficiary of the water diverted by all the dams on the controversial Narmada river system (Roy 1999). It needs this water for irrigation in an attempt to anchor the livelihoods of some desperately poor rural people in its hinterland who have not benefited from the state's economic growth. If it doesn't do something like this, they, too, will move to the cities such as Bhuj and Ahmedabad. But rural crisis and exodus have long been the by-products of 'modernization' and 'growth' in India. The original green revolution of the mid-1960s introduced hybrid seeds, and packages of practices and inputs for their use, that pushed small farmers off the land, which was accumulated by larger farmers. Farm production grew impressively, but the winners were urban people with purchasing power, not the dislocated rural poor (Shiva 1989; Dreze 1990).

Another 'push' from the countryside into these towns – this perhaps more speculative, but based on inference from other border situations – is that Indian army defensive operations near the border with Pakistan may have disrupted already fragile livelihoods. Finally, one must consider the cultural as well as economic and political situation of the tribal (*adavasi*) people in India generally and in Gujarat in particular. Many of the people who may have been losers and not winners in the growth stimulated by globalization, especially in the isolated northern parts of Gujarat, where the shaking was most extreme, are ethnic minorities. They have received very few government services over the years, tend to be displaced by so-called development projects, and often end up among the poorest of urban squatters when they are displaced.

## Alternatives

### *Nasty, brutish and short: is there an alternative?*

What alternatives are there? First, one can try to work within the existing capitalist order, showing corporations where their interests overlap with those of the general public. This is the realm of 'corporate responsibility' and marginal reforms (Twigg 2001). One can also try to work with marginal economic groups using innovative tools such as micro-credit for home and community safety improvements. This is similar to the approach of FEMA's Project Impact, the Inter-American Development Bank, and some NGOs in Latin America and India, as well as the approach in much of Europe, Australia and New Zealand, and the urban–industrial enclaves of Asia and the Pacific (IDB 2000). It is the approach that underlies the creation at the end of the International Decade for Natural Disaster Reduction (IDNDR) of the ProVention Consortium by the World Bank. ProVention's concerns range from novel and innovative 'catastrophe bonds' to assist the worldwide reinsurance industry to the use of 'micro credit' among the poor as a way of financing investments in home safety. Successes can be achieved and some improvements in public safety can result (World Bank 2001).

However, in the long run a second, complementary, initiative is necessary. Full hazard mitigation, as a mainstream part of sustainable development, is impossible without challenging the prevailing ideals of limitless growth, of ever-decreasing governmental regulation, and of the dominance of market values. The broad coalition of unions, environmental groups, consumer advocates, churches and other citizen-based organizations present in Seattle, Prague and Genoa to protest the World Trade Organization show that such questions are being asked widely. Education of citizens and workers concerning the interactions of power, economics, land use, well being and safety provides support for those on boards of directors and those in government that realize that 'sustainable growth' is an impossibility. Thus an important strategic move would be to link hazard mitigation efforts to the agendas of unions and citizen-based groups concerned with electoral reform, corporate accountability and environmental justice. A good start would be for such popular coalitions to push for a national dialogue on sustainable development as the context for recovery planning whenever a disaster occurs. Following Hurricane Mitch, a broad cross-section of Nicaragua's citizens demanded such a dialogue. In the post-Mitch period, civil society in Nicaragua succeeded in articulating a new vision of sustainable development in that country

(Barraclough and Moss 1999; cf. Bommer 1985). For so-called 'recovery' to be anything more than the re-establishment of the *status quo ante* that makes people, schools, hospitals vulnerable in first place, there must be a broader development vision. Business as usual will only reproduce the pre-conditions for yet more disasters (Susman *et al.* 1983; Blaikie *et al.* 1994; Hewitt 1996).

This, then, brings us full circle, back to remarks at the beginning that one needs to look beyond the 'risk society' at the relations of power and material interest that create risk. Globalization is making the game of life more risky for many as it provides an illusion of security for some (although the illusion was severely tested by the events of 11 September 2001). The social and spatial distribution of risks of all kinds are changing. We must make every effort to understand these changes and to work in solidarity with civil society to demand alternatives (Bernard and Young 1997; Pye-Smith *et al.* 1994; Varley 1994; Maskrey 1989). The reasons, to quote two fine Western authors: 'No man is an island' (John Donne), and 'It is not enough to understand the world, one must work to change it' (Karl Marx).

### *A human rights approach to disaster prevention?*

More and more people affected by disasters such as those mentioned here – Hurricane Mitch, earthquakes in Armenia, El Salvador, Gujarat – are aware that their suffering could have been avoided. There is pressure for new laws, for enforcement of old ones. Such a change in consciousness may be translated into a demand for recognition of the human right to protection from avoidable harm in extreme natural events (Handmer 2000; Kent 2000; Wisner 2000). The foundations for such an approach have already been laid by the work of international organizations and human rights activists working on other questions (Johnston 1997; Aysan 2000; Boyce 2000; IFRC 2000: 145–157; Molin 2000). During the International Decade for Natural Disaster Reduction (IDNDR), which ran from 1990–1999, three kinds of progress was made. Together they are necessary, but not sufficient, to initiate a sea change in how nations deal with natural hazards. These areas of progress concerned diffusion of technical knowledge, support for institution building, and financial assistance. The missing ingredient during the IDNDR was a moral imperative that can mobilize political will. It is when the world at large agrees to standards of responsibility by nation-states towards their citizens in the form of treaties, covenants and other agreements, that this moral force is felt most strongly.

Why, then, not set our sights on an international treaty that commits governments around the world to apply low-cost solutions based on available knowledge to prevent such tragic, avoidable loss?

Networks of scientists and engineers exist that could take on the technical work of defining these standards. These networks were created in part by the IDNDR. But there was unfinished business. Science was exchanged, but generally it hasn't been applied. Such an effort would require thousands of experts to work out the low-cost, minimum practices required to avoid further such tragedies. These scientists and engineers would have to sit down with lawyers, legislators and policy experts to work out how the minimum standards would be enforced. This is not an impossible task. It has happened before. One recent example is exchange among hundreds of agencies that work in humanitarian and disaster relief that led to agreement on a very detailed set of minimum technical standards for relief. Known as the SPHERE project (SPHERE 2000), its published document covers food, water, shelter, healthcare and many other aspects of relief. There are also many internationally agreed safety standards for the chemical industry, airline industry, nuclear power industry, etc. It has happened already where global warming is concerned. The Inter-governmental Panel on Climate Change (IPCC) has mobilized thousands of scientists, and their work has gone into the treaty-making process that led to the Kyoto Accord on greenhouse gas emissions.

Could the UN not create a parallel Inter-governmental Panel on Natural Disaster, that would, in a similar way, act to mobilize existing knowledge and feed it into a treaty-making process (Burton 2000)? Such a body is necessary because so many different kinds of knowledge and expertise is required. No single existing specialized agency of the UN, such as UNESCO, UNEP, WHO or WMO, covers all the specialist knowledge that would be required. That is one of the reasons that the IPCC was created. Preparing for the impacts of global warming requires many kinds of knowledge from areas such as public health, economics, agriculture, oceanography, in addition to expert understanding of world and regional climate.

What would be done during the many years that such a treaty would be in the making? The beauty of this process is that the low-cost solutions will filter out into society. Citizen groups will continue to demand action by their governments. They did so in Turkey when it became clear that contractors hadn't followed building codes and had used low-quality materials. They also demanded change in South Florida, in the USA, when it came to light that poor construction methods were responsible for much avoidable damage in Hurricane Andrew. Prevention of disasters has to come from the bottom up as well as from the top down (Kirby *et al.* 1995; Wisner 1993a, 1993b, 1995; FEMA 1997; Platt 1999; Alexander 2000; Harvey 2000).

## Conclusion

Absolute safety is not a human right. Safety from avoidable loss, injury and death is. Nothing in the Universal Declaration of Human Rights makes much sense if the human beings who are supposed to enjoy these rights can be snuffed out because a government neglected to enforce its own building codes. An international campaign in favour of establishing this human right as a part of international law would make the next step in the growing awareness by ordinary people that distant decisions in the global economy create risks that affect their daily lives.

## References


Abramovitz, J. (2001) 'Averting unnatural disasters', in L. Brown, C. Flavin, and H. French (eds) *State of the World 2001*, New York: W.W. Norton.

Alexander, D. (2000) *Confronting Catastrophe: New Perspectives on Natural Disasters*, New York: Oxford University Press.

Ayres, R. and Weaver, P. (eds) (1998) *Eco-Restructuring: Implications for Sustainable Development*, Tokyo: United Nations University Press.

Aysan, Y. (2000) '"Putting floors under the vulnerable": disaster reduction as a strategy to reduce poverty', <http://www.anglia.ac.uk/geography/radix/humanrights2.htm> (accessed 24 February 2002).

Bales, K. (1999) *Disposable People: New Slavery in the Global Economy*, Berkeley: University of California Press.

Barraclough, S. and Moss, D. (1999) *Toward Greater Food Security in Central America Following Hurricane Mitch: Rethinking Sustainable Development Priorities*, Boston: Oxfam America.

Beck, U. (1992) *Risk Society: Towards a New Modernity*, London: Sage.

Bernard, T. and Young, Y. (1997) *The Ecology of Hope: Communities Collaborate for Sustainability*, Gabriola Island, BC, Canada: New Society Publishers.

Blaikie, P., Cannon, T., Davis, I. and Wisner, B. (1994) *At Risk: Natural Hazards, People's Vulnerability, and Disasters*, London: Routledge.

Bommer, J. (1985) 'The politics of disaster – Nicaragua', *Disasters* 9 (4), 270–278.

Boyce, J.K. (2000) 'Let them eat risk? Wealth, rights and disaster vulnerability', *Disasters* 24 (3), 254–261, <http://www.anglia.ac.uk/geography/radix/humanrights.htm> (accessed 15 January 2002).

Burby, R. (ed.) (1998) *Cooperating with Nature: Confronting Natural Hazards with Land-Use Planning for Sustainable Communities*, Washington, D.C.: Joseph Henry Press.

Burton, I. (2000) 'The Intergovernmental Panel on Natural Disasters', <http://www.apu.ac.uk/geography/radix/standards.htm#Ian.Burton> (accessed 8 March 2002).

Chossudovsky, M. (1997) *The Globalization of Poverty*, London: Zed Books.

Comfort, L. (1999) *Shared Risk: Complex Systems in Seismic Response*, Amsterdam: Pergamon.

Comfort, L. *et al.* (1999) 'Reframing disaster policy: the global evolution of vulnerable communities', *Environmental Hazards* 1, 39–44.

Corbridge, S., Martin, R. and Thrift, N. (eds) (1994) *Money, Power and Space*, Oxford: Blackwell.

Cutter, S. (1993) *Living with Risk: The Geography of Technological Hazards*, London: Edward Arnold.

Daly, H. and Cobb, J. (1989) *For the Common Good: Redirecting the Economy Toward Community, the Environment, and a Sustainable Future*, Boston: Beacon.

Diaz, H. and Pulwarty, R. (eds) (1997) *Hurricanes: Climate and Socio-Economic Impacts*, Berlin: Springer-Verlag.

Dreze, J. (1990) 'Famine prevention in India', in J. Dreze and A. Sen, *The Political Economy of Hunger*, Volume 2, Oxford: Clarendon.

Durning, A. (1992) *How Much is Enough? The Consumer Society and the Future of the Earth*, New York: W.W. Norton.

Enarson, E. and Morrow, B. (eds) (1997) *The Gendered Terrain of Disaster: Through Women's Eyes*, Westport, Conn.: Praeger.

Federal Emergency Management Agency (FEMA) (1997) *Multi-Hazard Identification and Risk Assessment*, Washington, D.C.: FEMA.

Godschalk, D., Beatley, T., Berke, P., Brower, D.J. and Kaiser, E.J. (1999) *Natural Hazard Mitigation: Recasting Disaster Policy and Planning*, Washington, D.C.: Island Press.

Handmer, J. (2000) 'Human rights and disasters: does a rights approach reduce vulnerability?', <http://www.anglia.ac.uk/geography/radix/humanrights5.htm> (accessed 21 December 2001).

Harvey, D. (1982) *The Limits to Capital*, Chicago: University of Chicago Press.

Harvey, D. (1996) *Justice, Nature & the Geography of Difference*, Oxford: Blackwell.

Harvey, D. (2000) *Spaces of Hope*, Berkeley: University of California Press.

Harvey, D. (2001) *Spaces of Capital: Toward a Critical Geography*, London: Routledge.

Hewitt, K. (ed.) (1983) *Interpretations of Calamity*, Boston: Allen and Unwin.

Hewitt, K. (1996) *Regions of Risk*, London: Longman.

Inoguchi, T., Newman, E. and Paoletto, G. (eds) (1999) *Cities and the Environment: New Approaches for Eco-Societies*, Tokyo: United Nations University Press.

Inter-American Development Bank (IDB) (2000) *Facing the Challenge of Natural Disasters in Latin America and the Caribbean: An IDB Action Plan*, Washington, D.C.: Inter-American Development Bank, <http://www/iadb.org/sds/env> (accessed 14 November 2001).

International Federation of Red Cross and Red Crescent Societies (IFRC) (2000) *World Disaster Report 2000*, Geneva: IFRC.

International Federation of Red Cross and Red Crescent Societies (IFRC) (2001) *World Disaster Report 2001*, Geneva: IFRC.

Jaeger, C. (1994) *Taming the Dragon: Transforming Economic Institutions in the Face of Global Change*, Yverdon, Switzerland: Gordon and Breach Science Publishers.

Johnston, B. (ed.) (1997) *Life and Death Matters: Human Rights and the Environment at the End of the Millennium*, Walnut Creek, Calif.: Altamira Press.

Kent, G. (2000) 'The human right to disaaster mitigation and relief', <http://www.anglia.ac.uk/geography/radix/humanrights.htm> (accessed 4 December 2001).

Kirby, J., O'Keefe, P. and Timberlake, L. (eds) (1995) *The Earthscan Reader in Sustainable Development*, London: Earthscan Publications.

Korten, D. (1995) *When Corporations Rule the World*, West Hartford, Conn.: Kumarian Press.

Linnerooth-Bayer, J., Loestedt, R. and Sjoestedt, G. (2000) *Transboundary Risk Management*, London: Earthscan.

Mander, J. and Goldsmith, E. (eds) (1996) *The Case Against the Global Economy and for a Turn Towards the Local*, San Francisco: Sierra Club Books.

Maskrey, A. (1989) *Disaster Mitigation: A Community Based Approach*, Oxford: Oxfam.

Mileti, D. (1999) *Disasters by Design: A Reassessment of Natural Hazards in the United States*, Washington, D.C.: Joseph Henry Press.

Mitchell, J. (ed.) (1999) *Crucibles of Hazards: Mega-Cities and Disasters in Transition*, Tokyo: United Nations University Press.

Molin, H. (2000) 'Reduccion de Desastres como un Derecho Humano', <http://www.anglia.ac.uk/geography/radix/humanrights3.htm> (accessed 6 March 2002).

O'Connor, M. (ed.) (1994) *Is Capitalism Sustainable?*, New York: Guilford Press.

Peacock, W., Morrow, B. and Gladwin, H. (eds) (1997) *Hurricane Andrew and the Reshaping of Miami*, London: Routledge.

Perrow, C. (1984) *Normal Accidents: Living with High-Risk Technologies*, New York: Basic Books.

Platt, R. (1999) *Disasters and Democracy: The Politics of Extreme Natural Events*, Washington, D.C.: Island Press.

Pye-Smith, C., Feyerabend, G. and Sandbrook, R. (1994) *The Wealth of Communities*, West Hartford, Conn.: Kumarian Press.

Reed, D. (ed.) (1996) *Structural Adjustment, the Environment and Sustainable Development*, London: Earthscan.

Roy, A. (1999) *The Cost of Living*, London: Flamingo.

Rubio, R., Arriola, F.J. and Aguilar, J.V. (1997) *Crecimiento Esteril o Desarrollo*, San Salvador: FUNDE.

Sachs, W. (ed.) (1999) *Planet Dialectics*, London: Zed Books.

Sassen, S. (1998) *Globalization and its Discontents*, New York: The New Press.

Satterthwaite, D. (ed.) (1999) *The Earthscan Reader in Sustainable Cities*, London: Earthscan.

Schor, J. (1996) *The Overworked American*, Cambridge, Mass.: Harvard University Press.

Shiva, V. (1989) *Staying Alive: Women, Ecology and Development*, London: Zed.

Shrivastava, P. (1992) *Bhopal: Anatomy of a Crisis*, London: Paul Chapman Publishing.

SPHERE Project (2000) *Humanitarian Charter and Minimum Standards in Disaster Response*, Geneve: SPHERE Project, <http://www.sphereproject.org/> (accessed 23 September 2001).

Susman, P., Wisner, B. and O'Keefe, P. (1983) 'Global disasters: a radical perspective', in K. Hewitt (ed.) *Interpretations of Calamity*, Boston: Allen and Unwin.

Twigg, J. (2001) *Corporate Social Responsibility and Disaster Reduction: A Global Overview*, Unpublished report to DFID, London: Benfield Greig Hazard Research Centre, University of London, <http://www.bghrc.com/> (accessed 8 March 2002).

United Nations Development Programme (UNDP) (1990–2001) *Human Development Reports* [annual], New York: Oxford University Press.

United Nations Development Programme (UNDP) (forthcoming) *World Vulnerability Report*.

United Nations Research Institute for Social Development (UNRISD) (1995) *States of Disarray: The Social Effects of Globalization*, Geneva: UNRISD.

United Nations Research Institute for Social Development (UNRISD) (2000) *Visible Hands: Taking Responsibility for Social Development*, Geneva: UNRISD.

Varley, A. (ed.) (1994) *Disasters, Development and Environment*, Chichester: Wiley.

Verluise, P. (1989) *Armenia in Crisis: The 1988 Earthquake*, Detroit: Wayne State University Press.

Weizsaecker, E., Lovins, A. and Lovins, H. (1997) *Factor Four: Doubling Wealth – Halving Resource Use*, London: Earthscan.

Wisner, B. (1993a) 'Disaster vulnerability: scale, power, and daily life', *GeoJournal* 30 (2), 127–140.

Wisner, B. (1993b) 'Disaster vulnerability: geographical scale and existential reality', in H.-G. Bohle (ed.) *Worlds of Pain and Hunger*, Freiburg Studies in Development Geography 5, Saarbrücken and Fort Lauderdale (Fla.): Breitenbach Publishers.

Wisner, B. (1995) 'Bridging "expert" and "local" knowledge for counter-disaster planning in urban South Africa', *GeoJournal* 37 (3), 335–348.

Wisner, B. (1996) 'The geography of vulnerability', in J. Uitto and J. Schneider (eds) *Preparing for the Big One in Tokyo: Urban Earthquake Risk Management*, Tokyo: United Nations University.

Wisner, B. (1998) 'The geography of vulnerability: why the Tokyo homeless don't "count" in earthquake preparations', *Applied Geography* 18 (1), 25–34.

Wisner, B. (1999) 'There are worse things than earthquakes: hazard vulnerability and mitigation capacity in Greater Los Angeles', in K. Mitchell (ed.) *Crucibles of Hazard: Mega-Cities and Disasters in Transition*, Tokyo: United Nations University Press.

Wisner, B. (2000) 'Disasters: what the United Nations and its world can do', *United Nations Chronicle* 37 (4), 6–9, <http://www.un.org/Pubs/chronicle/2000/issue4/0400p6.htm> (accessed 8 March 2002).

Wisner, B. (2001) 'Risk and the neoliberal state: why post-Mitch lessons didn't reduce El Salvador's earthquake losses', *Disasters* 25 (3), 251–268.

World Bank (2001) *World Development Report 2000–2001*, Washington, D.C.: World Bank.

# 4 Gender, disaster and development

## The necessity for integration

*Maureen Fordham*

### Introduction

Gender, disaster and development have traditionally been treated as separate categories within academic disciplines and in terms of professional practice. Workers and researchers only rarely transfer knowledge between them and yet each group could contribute much to the others. However, it is increasingly being recognized that these elements must be brought together to improve understanding and practical action. When disasters occur, they bring with them a convergence of external help focused on providing immediate relief and quickly returning the community to 'normality'. In doing so, they can overturn long-term development programmes; the 'tyranny of the urgent' (BRIDGE 1996) can drive out gender and other fundamental social issues, or relegate them to a lower priority. Similarly, many development programmes are planned and undertaken without ensuring they do not exacerbate hazardous conditions or make people (and particularly women) more vulnerable to disasters.

The following discussion argues for the necessity of integrating gender, disaster and development in order to move from vulnerability to greater resilience in disaster-struck and disaster-prone areas. The building of sustainable, disaster-resistant communities, in both the developed 'North' and the developing 'South', requires increased employment of interdisciplinary knowledge and initiatives – although, for practical reasons, the weighting of the elements is rarely likely to be equal. Arguably, such a melding is more acceptable to those working in the development arena, where it is not so much a radically new suggestion (although somewhat elusive in practice), than in hazard and disaster management in developed countries, where it may be regarded as unnecessary or institutionally problematic. However, there is a need in the North for socially inclusive, participatory disaster management which can benefit from lessons learned in the South. This requires a reversal in the dominant direction of information flow: from South to North, not North to South.

Notwithstanding the call for integration, the difficulties of such a broad approach are fully acknowledged and the requirement to foreground one or another at different times is accepted. Indeed, the primary focus of this chapter is located within my original disciplinary 'home' of hazard and disaster management, rather than my adopted home of development. This background necessarily influences my approach, as it will be expected to do with others. The main aims of this chapter are to examine the character of the separated concepts (gender, disaster, development) and to point out some shortcomings of a discrete approach before addressing the challenges of integration.

I will not defend or discuss further the use of ideological terminology such as 'Third World', 'developed/developing world', 'North/South', etc. beyond noting its problematic nature and admitting that none of the terms used are wholly acceptable. Furthermore, the very distinction between developed and developing parts of the world is itself a challenging one when it must be accepted that there exist rich and poor, developed and less-developed areas, within both the North and the South. Rather, the relationship is a dialectical one that recognizes the colonial/post-colonial history of their yet-continuing interaction in the context of globalization.

## Hazard and disaster research: present and absent perspectives

We must begin with the recognition of a further fragmentation: that between hazard and disaster studies. Broadly defined, the concerns of hazards research have been claimed as:

> The totality of factors which generate, sustain, exacerbate, or mitigate those characteristics of natural and man-made environments [sic] that threaten human safety, emotional security, and material well-being.
>
> (Mitchell 1984: 37)

However, this claim to 'totality' is rarely substantiated in the literature. Geographers' (in particular) prior concern has been more with the physical hazard agent rather than disaster (i.e. the essentially *social* outcome of hazardous events); and more with individual hazard perception (or misperception) than underlying social structures, which create inequalities and vulnerability. Hazards research generally has drawn criticism for its lack of attention to social theory.

### *The dominant paradigm*

This early hazards work has been variously called 'a modest but universal research paradigm' (Kates and Burton 1986: 324) and, more critically, the 'dominant view' (Hewitt 1983: 4). Research workers in the hazards field have tended not to challenge or extend the fundamental theoretical base, but rather to make incremental changes within the paradigm. Quantitative measurement techniques have favoured the concentration of explanatory power at the individual unit of analysis, to the neglect of the political–economic context (similarly, concentration on the individual unit of analysis has favoured quantitative measurement techniques). This has led to accusations of banality and triviality (Watts 1983; Waddell 1977), an inappropriate emphasis on the individual, and a failure to deal with constraints on preference and choice (Fordham 1992).

Hewitt's *Interpretations of Calamity* (1983) provided a critical alternative to the dominant view, with its interpretation of disasters as phenomena outside the ordinary and the everyday:

> (a) Most natural disasters, or most damages in them, are *characteristic* rather than accidental features of the places and societies where they occur.
> (b) The risks, pressures, uncertainties that bear upon awareness of and preparedness for natural fluctuations flow mainly from what is called 'ordinary life', rather than from the rareness and scale of those fluctuations.
> (c) The natural extremes involved are, in a human ecological sense, more expected and knowable than many of the contemporary social developments that pervade everyday life.
>
> (Hewitt 1983: 25)

Hewitt's critique is something of a defining moment in hazards research as it brought together a sustained challenge to the dominant paradigm and marked, if not its overthrow, at least the possibility of a more diverse conceptual base. It also marked a necessary shift in thinking about appropriate *solutions* – away from the technical fix and towards political economy – and signalled a change in *process*, away from top-down, expert-led practice towards more participatory decision-making processes and implementation. Interestingly, in light of the concerns of this chapter, it drew many of its examples from development studies and the South, whereas the dominant paradigm was located more firmly in the developed North.

The most prominent and long-lasting, alternative perspective coming out of this critique has been the vulnerability approach (for discussions see Hewitt 1997; Blaikie *et al.* 1994; Varley 1994; Comfort *et al.* 1999; Cannon 1994, 2000). This has an acknowledged political stance which more 'traditional' hazards researchers have tried to refuse or downplay. The vulnerability perspective has as its major focus the 'social geography of harm' (Hewitt 1997); that is, on the underlying, socio-political, root causes of hazardous places and disaster processes, rather than physical hazard agents or superficial symptoms. Comfort *et al.* (1999) propose that:

> Human vulnerability – those circumstances that place people at risk while reducing their means of response or denying them available protection – becomes an integral concern in the development and evaluation of disaster

> policies. We must change the policies of today that rely heavily on sending assistance only after tragedy has occurred.
>
> (Comfort *et al.* 1999: 39)

An example that recognizes the differential social impacts of hazards and disasters, and extends 'disaster' into 'development', is that by La Red (the social studies network for disaster prevention in Latin America) which has compiled the DesInventar database identifying the impact of many local, small-scale 'events' in worsening vulnerability. These events may not meet the criteria for definition as a 'disaster' (usually referring to large-scale events which overwhelm local (official/administrative) coping abilities), but it is estimated that the cumulative impact of these small events is equal to or exceeds those for large-scale disasters. These small-scale events impact the poor more than the rich and do not attract the external assistance and resources that are common in large-scale disasters (Sequira 2001).

ITDG (2001) notes that most disaster management/relief efforts go to large-scale, national/international-level actors rather than to the neighbourhood/community level where women are most evident. Yet, experience of working with women after the 1999 Marmara earthquake in Turkey leads Sengul Akçar to state:

> [W]ithout effective, poverty reducing, community development approaches and without strengthening grassroots democracy and local self-governance, good/effective post-disaster response cannot be expected.
>
> (Akçar 2001: 3)

This kind of work indicates a considerable shift in emphasis that brings it more into line with the stated aims of this chapter. Handmer (2000: 278) describes some of the shifts and trends that have occurred in hazard research and management in terms of, *inter alia*, moving from a reactive hazards approach to one of proactive risk management in the context of vulnerability; from top-down expert planning to partnerships with those at risk; from unidimensional to integrated approaches incorporating economic, social and environmental issues with hazards. This, however, describes a positive view of change in the field and, despite many advances, we must recognize more rhetoric than reality in actual practice. Blaikie *et al.* (1994) argue that much disaster work treats symptoms rather than causes because:

> Vulnerability is deeply rooted, and any fundamental solutions involve political change, radical reform of the international economic system, and the development of public policy to protect rather than exploit people and nature.
>
> (Blaikie *et al.* 1994: 233)

It is this fundamental challenge to the status quo that most particularly splits the research field. Much resistance remains to a political economy/

vulnerability approach (Bryant 1991) and to a perceived stridency of expression which 'at worst, simply call[s] for overall social revolution' (Smith 1996: 51).

Where a willingness to challenge the existing social order is sometimes more acceptable is in the academic and practitioner fields of development studies and development respectively, although, in saying this, I am not denying the economistic (indeed, econometric) nature of much development theory. First, however, I must mention briefly the (broadly) sociological perspective on disasters (as opposed to hazards) which has also been dominated by North American scholarship and a Northern/Western focus. European scholars have also made significant contributions but have more recently favoured the risk area of research, with a similarly Northern/Western focus. Whilst this disciplinary background makes a greater contribution to understanding the social construction of disaster, it does so through an examination of collective behaviour and organizational analysis, particularly of civil defence emergency planning structures. It does not, in the main, embrace the radical political implications of a vulnerability perspective and retains – with certain exceptions – a gender-insensibility indicative of a masculine-dominated professional/practitioner and research environment. In a piece discussing both present and absent perspectives in disaster research, Hewitt (1998) asks rather bleakly:

> But where does organizational sociology lead when it takes improving the effectiveness and centralized administration of agencies and expert systems as its focus? Can it offer any advice, other than a more totalizing penetration of government and powerful interests into everyday life, a greater surveillance and militarization of public and private space?
>
> (Hewitt 1998: 90)

Quarantelli (1998: 260), in the same volume, notes the lack of congruence between both the topics and the literature used by scholars who all, nevertheless, call themselves disaster researchers, and surmises it may be a result of the North–South/disaster–development split with which this chapter is concerned. Whilst excluding, on theoretical grounds, famines, epidemics and droughts from the umbrella term 'disasters', he nonetheless recommends a closer examination by disaster researchers in the North of empirical data and theoretical ideas from the South. Such a shift has greater possibility of success in academic circles than it faces in practical emergency planning and disaster management, where personal experience suggests significant resistance to the adoption of ideas and practices emanating from different cultural milieux.

## Development and development studies

It is of course a considerable oversimplification to speak of the field of development studies as if it were both single and uncontested. Nevertheless, while acknowledging Stuart Corbridge's observation that 'development studies as an intellectual discipline has expanded so much in the past twenty or thirty years that it is in danger of losing its claim to distinctiveness' (1995: x), there is a sense in which there is a common, generalized understanding of what development studies is.

> If development studies means anything it presumably means the study of those countries which have a past in common, if not a common past – I refer to a history of colonialism and post-colonialism – and which can only reasonably be understood in relation to the so-called developed countries.
>
> (Corbridge 1995: x–xi)

Beyond this, development itself has been defined as 'an economic, social and political process which results in a cumulative rise in the perceived standard of living for an increasing proportion of a population' (Hodder 2000: 3). But development studies, as much as hazard and disaster studies, has a history and has also undergone changes in dominant and competing theoretical positions: from modernization theorists, through dependency, to anti-dependency theorists and others. But it is to work in the last ten or fifteen years or so that we can look for useful additions to a gendered development theory and practice. This work has highlighted the vulnerability and capacity of different social groups, especially women, and developed more participatory approaches to decision-making and policy implementation.

However, many development workers are not sensitized to the needs of disaster mitigation planning, and rigid bureaucratic structures (particularly related to finance and other resourcing issues) mitigate against such synergistic working. Thus there is much to be transferred between the development and disaster communities.

Anderson and Woodrow (1998) describe some of the tension between disaster and development aims:

> The commitment to development is clear and unshakable among non-governmental organizations. However, when circumstances require them to respond with immediate humanitarian relief, development goals are often lost or at least deferred while emergency efforts prevail . . . With regret, agencies feel that they cannot maintain their commitment to development while disaster response is demanded.
>
> (Anderson and Woodrow 1998: 7)

Similarly, research in South Sudan (Palmer 1998) found reproductive healthcare – more commonly a part of development programmes – had been undervalued by agencies involved in disaster and crisis situations (which tend to emphasize food, water, shelter and emergency medical care), even though it was identified as significant by community members themselves. The issue of reproductive health is a sensitive one. On the one hand, this important matter is often neglected or unrecognized by the dominantly male decision-makers and thus increases the negative impacts of disasters (and development) on women. On the other hand, women tend all too often to be defined (and confined) by their reproductive capabilities rather than – or in addition to – their productive capacities. It is a typical example of the familiar private–public, socio-spatial divide that confines women to the domestic sphere.

Anderson and Woodrow are amongst those to have outlined an analytical framework to help workers 'hold fast' to development aims during the disaster process. This uses a matrix to analyse vulnerabilities and capacities. The authors define development itself as 'the process by which vulnerabilities are reduced and capacities increased' (1998: 12). The recognition of the two sides is another important conceptual advance for disaster studies that comes out of the development field, although it was perhaps an idea whose 'time had come' as it coincided with similar ideas in social theory more generally. Three areas of vulnerability and capacity are represented: physical/material, social/organizational, and motivational/attitudinal. And in order to better signify complex reality, further dimensions are added which include disaggregation by gender and other differences (e.g. class; ethnic, political or language group; rural/urban; age; etc.).

While vulnerabilities analysis has been used relatively widely to examine susceptibility to hazard, it is used more broadly here to encompass positive, not just negative, aspects and, importantly, development potential. While leading-edge disaster research and practice in the developed world is beginning to recognize this broader concept of vulnerability, it has yet to grasp fully the potential of capacities analysis. Anderson and Woodrow argue: 'Capacities assessment is critical for designing projects to have a positive development impact' (1998: 14).

To emphasize the significance of capacity, the authors place this word first, and name the process 'Capacities and Vulnerabilities Analysis' (1998: 15). Such an analysis, if used in the developed world context, would radically change the process of disaster management. It would, *inter alia*, fundamentally challenge the dominant command and control model; build in longer-term aspects; treat disaster mitigation as integral to everyday social and economic development; and develop local community capacities to complement or replace 'official' services through the greater use of, and willingness to work with, NGOs. In the context of this discussion, it would recognize women as active agents and not simply as victims.

## Gender

Gender is a key dimension of social difference and yet it has been absent as an analytical variable in much development and disaster research (although less so in the former than the latter). Ester Boserup's *Woman's Role in Economic Development* was an early (1970) exception to this. Little appeared in the disaster literature, however, until, for example, Joan Rivers's (1982)

'Women and Children Last: An Essay on Sex Discrimination in Disasters', which identified the nutritional vulnerability of female children leading to higher incidences of malnutrition in girls in famines as a result of 'sex discrimination intrinsic in most societies' (Rivers 1982: 265). These both come from a 'development' background and focus on women specifically. Gender analysis came later to disaster research in the developed North. Alice Fothergill's review (in Enarson and Morrow 1998) of disaster research incorporating a gender analysis uncovers some 100 studies. However, in many of these, gender is simply a quantitatively measured background characteristic rather than a central analytical category.

Even into the 1990s, many books on hazards and disasters failed to recognize the significance of analytical categories such as 'gender', 'women' or 'feminism', to the extent that such terms do not warrant inclusion in their respective indexes, or even sometimes in the text as a whole. Thus disaster research – somewhat more so than development research – has been gender-insensitive. Men's experiences and conceptualizations have been used as a universal category, and in response to that much gender research has focused on women because of their relative invisibility and their, now demonstrated, greater potential vulnerability. Even in 2001 the UNDP Human Development Report reported that, compared to men, women's achievements lag behind and their deprivations are greater. However, women are not a homogeneous category and when speaking of women we must also recognize difference in terms of race/ethnicity, class/caste, sexuality, [dis]ability, etc. which intersect in complex ways with gender (Enarson and Fordham 2001; Fordham 1999, 1998; see also Bullard 1990 (in terms of environmental justice); Cutter *et al.* 1992; and Flynn *et al.* 1994 (in terms of risk perception); and Peacock *et al.* 1997 (addressing several of these aspects)). Many of these aspects of differentiated risk are set out in Box 4.1.

An area that has attracted greater attention in the development arena in recent years has been violence to women (Pickup *et al.* 2001); in the (Northern/Western) disaster management field it emerged rather late and on a small scale (Larabee 2000; Ralph 1999; Fothergill 1999; Wilson *et al.* 1998; Enarson 1997). Despite the recent, relative visibility, Pickup *et al.* contend that: 'Violence against women is probably still regarded as an important, but fairly marginal, concern by most staff in most international development and relief agencies' (2001: 301). Part of the reason for this is the dominance of men in the fields of disaster management and development (perhaps more so in the former compared to the latter), and the frequent invisibility of gendered violence issues in their associated masculine, often militaristic, cultures (Enarson and Fordham 2001).

Pickup *et al.* (2001) recommend organizations to focus on three forms of policy implementation, familiar to other gender analyses: mainstreaming awareness of violence against women into all development, relief, and advocacy interventions; developing an integrated organizational strategy for interventions; and using broad and strategic alliances. However, their recommendations acknowledge the often hidden risks of such interventions in

**Box 4.1 Women at risk**

- Poor or low-income women
- Refugee women and the homeless
- Senior women
- Women with cognitive or physical disabilities
- Women heading households
- Widows and frail elderly women
- Indigenous women
- Recent migrants
- Women with language barriers
- Women in subordinated cultural groups
- Socially isolated women
- Caregivers with numerous dependants
- Women in shelters/homeless
- Women subject to assault or abuse
- Women living alone
- Chronically ill women
- Undocumented women
- Malnourished women and girls

Source: Enarson (2000).

Some of the key factors which contribute to women's vulnerability in the South Asian context:

- Very high illiteracy levels
- Low ownership of assets (e.g. land)
- Minimum work opportunities outside the home
- Limited mobility outside the home and locality
- Low social status
- Socially constructed dependency on male relatives

Source: Ariyabandu (2000).

the form of backlash and reprisal – from both women and men who stand to gain from maintaining the status quo. Declining resources and increasing competition may impact on the perceived relevance or priority of initiatives on mainstreaming gender and/or violence against women. Furthermore, despite the numerous UN resolutions and agreements requiring gender mainstreaming in policies, programmes and institutions (UNDP 1998), women and women's voices are largely absent in policy-making institutions (Çağatay 2001: 9).

## Towards integration

Notwithstanding the important role of vulnerability in women's lives they also have great capacities to resist and overcome socially constructed disaster impacts, and recognition of this, too, has been a contribution to disaster research from the development field. For example, Ariyabandu (in Fernando and Fernando 1997) presents an 'alternative perspective' which combines gender, disaster and development, arguing that:

> The Alternative Perspective suggests that the differential impact of disasters on women is a manifestation of the failure of organisations involved in development and disaster mitigation, to understand that it is the structures and relationships that form gender relations in society which turn women into victims. At the same time, there is a wealth of information and experience which shows that within these victimising relationships women manage crisis situations, and demonstrate high degrees of resilience.
>
> (Ariyabandu 1997: 6)

Other examples of integrating programmes exist and four of these are presented below (see Boxes 4.2 to 4.5). Many of these have been spurred into action by failures in traditional practices.

### Box 4.2 Community exchanges

Reconstruction following the Latur earthquake in Maharashtra, India, in 1993 was initially managed by the men who failed to plan for water storage, grain storage, cooking or keeping of cattle. The men also preferred to demolish the old stone houses and rebuild in costly reinforced concrete. The development organization SSP (Swayam Shikshan Prayog – self-education for empowerment) facilitated the building of low-cost, safer structures in traditional brick or stone by local women. This knowledge was subsequently transferred to the women of Gujarat in a process which began as a show of solidarity between the women of these disaster-hit locations and ended as a long-term co-operation for the exchange of skills. Such community-to-community exchanges are seen by SSP as a tool for horizontal learning and transfer of ideas (SSP-India 2001).

**Box 4.3 Gender relations**

Even apparently participatory, locally based interventions can have unforeseen negativities for women if development ignores the dynamics of gender relations (see Agarwal 1997; BRIDGE n.d.; Fordham 2001). Too often 'communities' are treated as ungendered units and 'community participation' as an unambiguous step towards enhanced equality; assumptions are made about intra-household equity; and the dangers of a male backlash are unrecognized. Increased power sharing at the local level may not automatically translate into power for women (Lind, cited in Cornwell 2000: 12). Development (and other) initiatives cannot be seen as static events but rather as dynamic processes of social relations.

**Box 4.4 Disaster preparedness**

After the 1991 cyclone in Bangladesh, that killed almost 140,000 people, a study carried out by the Red Crescent and other organizations found that 90 per cent of the victims were women and children. Despite the existence of cyclone shelters, the communities living near them were unaware of their purpose or felt unable to take refuge in them. The Community Based Disaster Preparedness Programme in Cox's Bazaar, Bangladesh (Schmuck 2002), which began in 1996, has the involvement of women as a major aim, although it is not easy to put into practice. The CBDPP tries to reduce the impact of cyclones on women and children and to empower them in their daily lives through a range of disaster management and development interventions, including group training sessions on disaster preparedness, leadership, reproductive health and nursery management techniques, as well as small-scale entrepreneurship such as chicken breeding. The programme also targets men by holding awareness sessions on gender equality during committee meetings.

Calls for integration are easy to make, but integration is difficult to achieve on the ground. Each of the concepts we have been examining is complex and problematic in various ways and, for practical reasons, the weighting of the different elements is rarely likely to be equal but always to be context-dependent. The building of sustainable, disaster-resistant communities, in

**Box 4.5 Mobilizing women's participation**

SSP (Swayam Shikshan Prayog – meaning self education for empowerment) is a development organization working towards sustainable development but, importantly for the focus of this chapter, it also works as a facilitator in transforming crisis situations (e.g. earthquakes) into opportunities to mobilize communities, and especially women (see Gopalan 2001; SSP-India 2001; *Housing by People in Asia* 2001: 3). SSP underlines the importance of building social infrastructure in the form of community resource centres which represent multiple benefits, including on-site training in earthquake-resistant construction and the building of community assets by women's groups. Following the 26 January 2001 earthquake in Gujarat, grassroots women's groups in 300 villages were empowered by the government to monitor progress in reconstruction. This community self-monitoring system was facilitated by SSP which noted several benefits, including:

- women's groups were involved in designing houses suitable to their needs;
- increased awareness and use of earthquake resistant technology;
- resource allocations were monitored, ensuring the accountability of implementing agencies and government officials.

Rehabilitation was dovetailed with local development in local communities (SSP Information sheet November 2001; Gopalan pers. comm. November 2001).

both the industrialized 'North' and the industrializing 'South', is a complex endeavour, the solution to which cannot easily be prescribed. Nevertheless, integrating gender, disaster and development, through interdisciplinary knowledge and initiatives in both the academic and practitioner communities, is required in order to improve the potential of moving from vulnerability to resilience in disaster-struck and disaster-prone areas.

## Gender, disaster and development in a globalizing world

Issues of gender, disaster and development must all be set within the context of a globalizing world that brings together complex processes operating at a range of scales.

Such emergent factors as market liberalization, capital mobility, increasing 'flexibility' in work practices and structural adjustment policies have had unequal and contradictory impacts, both geographically and socially. Increases in household income through enhanced production for export has been a positive benefit for some, but may not translate into gains in well-being for women and children (Çağatay 2001; Mies 1999: 81). While it is acknowledged that these global pressures also impact on men, their impacts on women often remain hidden. If women's labour is diverted from food crop production to cash crop production, family nutritional standards can be compromised (Çağatay 2001: 7), women's agricultural base can be undermined and rates of women's migration increased. In these circumstances, rural women are lured to unsafe living conditions and informal sector work in the increasing number of mega-cities where they are increasingly exposed to urban environmental pollution and disasters such as mudslides and earthquakes (Enarson 2000). The potential for an increased share of paid employment for women must be seen in the context of the erosion of workers' power deriving from intensified globalization and market liberalization. And so, for those women who gain from access to paid employment, their 'comparative advantage' (in economistic terms) comes from their lower wages and inferior working conditions (Çağatay 2001: 7). 'It is not simply that some women lose while others gain from trade expansion. Rather, as gender inequalities are multidimensional, even women who may gain in one dimension, such as employment, may lose in another, such as leisure time' (Çağatay 2001: 8). Trade expansion and trade liberalization have contradictory effects on women's well-being and gender relations, but 'the phenomenon of globalization . . . has so far proven to be a brutal one for many' (ibid.: 35).

The UK Disasters Emergency Committee (DEC) reflects something of the ambiguous impacts of globalization in its report on the earthquake in Gujarat, India, in 2001. They point to the substantial roles and freedoms given to civil society and the private sector (increasingly involving women) by the government of India, which previously had kept a firm control on emergency response. DEC argues that while this could be regarded positively as evidence of a new 'coherence', it could also be seen negatively as an abnegation of the responsibilities of government and hence an erosion of citizens' rights. This is a reflection of the global phenomenon of adjustment between the roles of the state, the private sector and civil society (ReliefWeb 2001). Furthermore, in terms of the fluidity of women's roles and positions, a gendered analysis points to adjustment *within*, not just *between*, these components.

## Conclusions

In arguing for the necessity of integration I am not underestimating the likely difficulties and the considerable challenge this represents for the future (Parker 2000). Institutional fragmentation, competition and misunderstanding are a major threat in both the academic and practitioner fields. One of the chief dangers is that the rhetoric of integration and inclusion masks a largely technocentric concern; one that may advance incrementally but still fails to grasp the difficult challenge of the root causes of vulnerability at the social and the political level. These represent a more far-reaching disturbance of the status quo and can trigger refusals to countenance such fundamental changes. Nevertheless, radical change (in varying degrees) is possible at both the abstract and the concrete level and is at the heart of transformative feminist practice. In the disaster and development context, Blaikie *et al.* (1994) address both the abstract and the concrete levels.

For academics and practitioners, the integration of gender, disaster and development requires a heightened awareness of ways of thinking and working that may have become more commonplace in one gender–disaster–development component rather than another. Integration has to operate at different levels:

1 The academic, theoretical level.
2 The policy level.
3 The practitioner level.
4 The political level.

Each of these must incorporate a degree of gender, disaster and development awareness that has become internalized in everyday practice. Sensitivity to social (Twigg and Bhatt 1998) and environmental (Leach *et al.* 1995) settings is imperative in transposing ideas and practices between developed and developing countries if methods are to be negotiable. A sense of total irrelevance can be conveyed to researchers/workers in these two domains by the inappropriate use of commonplace imagery and descriptions.

While a degree of resistance to political economy approaches has already been noted, there appears to be a more generalized resistance in the North to learning from the South. Development theorists and practitioners working in the South have formulated innovative strategies and methodologies that have potential value to workers/researchers in the North. Participatory approaches, which go beyond token consultative practices, have grown out of recognition of the shortcomings in top-down, expert outsider systems. Such systems have underplayed the social and environmental impacts of projects and programmes; ignored pre-existing networks and knowledge, better suited to local conditions; and undermined local economic structures. Furthermore, they reinforce the dominant flow of expertise from North to South.

Finally, I do not intend to provide a neat summing up and set of prescribed recommendations for the integration of gender, disaster and development. Despite a number of examples, I suggest we don't fully understand how to do this satisfactorily and it is unlikely that 'one size fits all'. Moreover, the first question for many is likely to be not whether or how such integration is possible, but whether it can be agreed that it is even desirable.

Researchers and practitioners in each of the fields and their intersections are at different points on the learning curve. Furthermore, the process of change is clearly not confined to a simple exchange of information but requires political will and, for some, a change in world-view. Despite the somewhat positive picture above of development's theoretical advances in the area of integration, it may be argued that the incorporation of a socially aware, gendered perspective exists largely at the level of rhetoric in the development studies/developing world context. But, I suggest, it has yet to reach even that point in practical disaster management in the developed world where the command and control model is dominant (Home Office 1998). Openness to what different disciplinary/professional perspectives have to offer is a beginning.

## References

Agarwal, B. (1997) 'Re-sounding the alert – gender, resources and community action', *World Development* 25 (9), 1373–1380.

Akçar, S. (2001) 'Grassroots women's collectives – roles in post-disaster efforts: potential for sustainable partnership and good governance', Paper presented at the UN Division for the Advancement of Women Expert Group Meeting on Environmental Management and the Mitigation of Natural Disasters: A Gender Perspective, 6–9 November 2001.

Anderson, M.B. and Woodrow, P.J. (1998) *Rising From the Ashes*, London: Intermediate Technology.

Ariyabandu, M.M. (1997) 'Preface', in P. Fernando and V. Fernando, *South Asian Women: Facing Disasters, Securing Life*, Colombo, Sri Lanka: Duryog Nivaran and Intermediate Technology Group.

Ariyabandu, M.M. (2000) 'Impact of hazards on women and children: situation in South Asia', Paper presented at Reaching Women and Children in Disasters, Miami, Florida, 4–6 June 2000.

Blaikie, P., Cannon, T., Davis, I. and Wisner, B. (1994) *At Risk: Natural Hazards, People's Vulnerability, and Disasters*, London: Routledge.

Boserup, E. (1970) *Woman's Role in Economic Development*, New York: St Martin's Press.

BRIDGE (1996) *Briefings on Development and Gender*, Issue 4: *Integrating Gender Into Emergency Response*, <http://ids.ac.uk/bridge/dgb4.html> (accessed April 2002).

BRIDGE (n.d.) *Coping with Conflict: The Case of Redd Barna Uganda*, Issue 9: *Gender and Participation*, <http://www.ids.ac.uk/bridge/dgb9.html#article3> (accessed April 2002).

Bryant, E.A. (1991) *Natural Hazards*, Cambridge: Cambridge University Press.

Bullard, R.D. (1990) *Dumping in Dixie: Race, Class, and Environmental Quality*, Boulder, Colo.: Westview Press.

Çağatay, N. (2001) *Trade, Gender and Poverty*, New York: United Nations Development Programme.

Cannon, T. (1994) 'Vulnerability analysis and the explanation of "natural" disasters', in A. Varley (ed.) *Disasters, Development and the Environment*, London: Belhaven Press.

Cannon, T. (2000) 'Vulnerability analysis and disasters', in D. Parker (ed.) *Floods*, London: Routledge.

Comfort, L., Wisner, B., Cutter, S., Pulwarty, R., Hewitt, K., Oliver-Smith, A., Weiner, J., Fordham, M., Peacock, W. and Krimgold, F. (1999) 'Reframing disaster policy: the global evolution of vulnerable communities', *Environmental Hazards* 1, 39–44.

Corbridge, S. (1995) *Development Studies: A Reader*, London: Arnold.

Cornwell, A. (2000) 'Making a difference? Gender and participatory development', IDS Discussion Paper No. 378, Institute of Development Studies, University of Sussex.

Cutter, S., Tiefenbacher, J. and Solecki, W.D. (1992) 'En-gendered fears: femininity and technological risk perception', *Industrial Crisis Quarterly* 6, 5–22.

Enarson, E. (1997) *Responding to Domestic Violence and Disaster: Guidelines for Women's Services and Disaster Practitioners*, Available from BC Institute Against Family Violence, 409 Granville, Ste. 551, Vancouver, BC, Canada V6C 1T2.

Enarson, E. (2000) *Gender and Natural Disasters*, Working Paper No. 1, InFocus Programme on Crisis Response and Reconstruction, Geneva: ILO, Recovery and Reconstruction Department.

Enarson, E. and Fordham, M. (2001) 'Lines that divide, ties that bind: race, class and gender in women's flood recovery in the US and UK', *Australian Journal of Emergency Management* 15 (4), 43–52.

Enarson, E. and Morrow, B.H. (eds) (1998) *The Gendered Terrain of Disaster: Through Women's Eyes*, Westport, Conn.: Praeger.

Fernando, P. and Fernando, V. (1997) *South Asian Women: Facing Disasters, Securing Life*, Colombo, Sri Lanka: Duryog Nivaran and Intermediate Technology Group.

Flynn, J. Slovic, P. and Mertz, C.L. (1994) 'Gender, race, and perception of environmental health risks'; *Risk Analysis*, 14 (6), 1101–1108.

Fordham, M. (1992) 'Choice and constraint in flood hazard mitigation: the environmental attitudes of floodplain residents and engineer', Unpublished Ph.D. thesis, Middlesex University.

Fordham, M. (1998) 'Making women visible in disasters: problematising the private domain', *Disasters* 22 (2), 126–143.

Fordham, M. (1999) 'The intersection of gender and social class in disaster: balancing resilience and vulnerability', *International Journal of Mass Emergencies and Disasters* 17 (1), 15–36.

Fordham, M. (2001) 'Challenging boundaries: a gender perspective on early warning in disaster and environmental management', Invited paper for United Nations Expert Group Meeting: 'Environmental Management and the Mitigation of Matural Disasters: A Gender Perspective', Ankara, Turkey, 6–9 November 2001.

Fothergill, A. (1998) 'The neglect of gender in disaster work: an overview of the literature', in E. Enarson and B. Morrow (eds) *The Gendered Terrain of Disaster: Through Women's Eyes*, Westport, Conn.: Praeger.

Fothergill, A. (1999) 'An exploratory study of woman battering in the Grand Forks flood disaster: implications for community responses and policies', *International Journal of Mass Emergencies and Disasters* 17 (1), 79–98.

Gopalan, P. (2001) 'Responding to earthquakes: people's participation in reconstruction and rehabilitation', Invited paper for United Nations Expert Group Meeting: 'Environmental Management and the Mitigation of Natural Disasters: A Gender Perspective', Ankara, Turkey, 6–9 November 2001.

Handmer J.W. (2000) 'Flood hazard and sustainable development', in D.J. Parker (ed.) *Floods*, London: Routledge.

Hewitt, K. (1983) *Interpretations of Calamity*, London: Allen and Unwin.

Hewitt, K. (1997) *Regions of Risk*, Harlow, Essex: Addison-Wesley Longman.

Hewitt, K. (1998) 'Excluded perspectives in the social construction of disaster', in E.L. Quarantelli (ed.) *What is a Disaster?*, London: Routledge.

Hodder, R. (2000) *Development in Geography*, London: Routledge.

Home Office (1998) *Dealing With Disaster* (3rd edn), London: HMSO.

*Housing by People in Asia* (2001) No. 13, <www.sspindia.org> (accessed May 2002).

Intermediate Technology Development Group (ITDG) (2001) 'Disaster reduction', <http:www.itdg.org> (accessed May 2002).

Kates, R.W. and Burton, I. (eds) (1986) *Geography, Resources and Environment: Selected Writings of Gilbert F. White*, Volume 2, Chicago: University of Chicago Press.

Larabee, A. (2000) *Decade of Disaster*, Urbana and Chicago, Ill.: University of Illinois Press.

La Red [Social Studies Network for Disaster Prevention in Latin America] (2002) <http://www.desinventar.org> (accessed May 2002).

Leach, M., Joekes, S. and Green, C. (1995) 'Editorial: gender relations and environmental change', *IDS Bulletin* 26 (1), 1–8.

Mies, M. (1999) 'Decolonizing the iceberg economy: new feminist concepts for a sustainable society', in L. Christiansen-Ruffman (ed.) *Social Knowledge: Heritage, Challenges, Perspectives. The Global Feminist Enlightenment: Women and Social Knowledge*, Pre-Congress volume. Madrid: International Sociological Association.

Mitchell, J.K. (1984) 'Hazard perception studies: convergent concerns and divergent approaches during the past decade', in T.F. Saarinen, D. Seamon and J.L. Sell, *Environmental Perception and Behavior: An Inventory and Prospect*, Research Paper No. 209, Chicago: University of Chicago, Department of Geography.

Palmer, C. (1998) 'Sudan research says listen to what people really want in emergencies', *Links, Oxfam Newsletter* No. 3.

Parker, D.J. (ed.) (2000) *Floods*, London: Routledge.

Peacock, W.G., Morrow, B.H. and Gladwin, H. (eds) (1997) *Hurricane Andrew: Ethnicity, Gender and the Sociology of Disasters*, London: Routledge.

Pickup, F., Williams, S. and Sweetman, C. (2001) *Ending Violence Against Women*, Oxford: Oxfam Publishing.

Quarantelli, E.L. (1998) *What is a Disaster?*, London: Routledge.

Ralph, R. (1999) 'Editorial: mapping gender violence', *ODI Relief and Rehabilitation Network Newsletter* No. 14.

ReliefWeb (2001) 'Disasters Emergency Committee', <http://www.reliefweb.int/w/rwb.nsf/s/2A50BDF2F1D8A377C1256AA9004B6905> (accessed May 2002).

Rivers, J.P.W. (1982) 'Women and children last: an essay on sex discrimination in disasters', *Disasters* 6, 256–267.

Schmuck, H. (2002) 'Empowering women in Bangladesh', *ReliefWeb*, <http://www.reliefweb.int/w/rwb.nsf/9ca65951ee22658ec125663300408599/570056eb0ae62524c1256b6b00587224?OpenDocument> (accessed May 2002).

Sequira, N. (2001) 'Risk management: an alternative perspective in gender analysis', Invited paper for United Nations Expert Group Meeting: 'Environmental Management and the Mitigation of Natural Disasters: A Gender Perspective', Ankara, Turkey, 6–9 November 2001.

Smith, K. (1996) *Environmental Hazards*, London: Routledge.

SSP-India (2001) *Lessons from the Epicentre: Mainstreaming Women's Initiatives in Disaster and Development*, <http://www.ssp-india.org> (accessed May 2002).

Twigg, J. and Bhatt, M. (1998) *Understanding Vulnerability: South Asian Perspectives*, Colombo, Sri Lanka: Duryog Nivaran and Intermediate Technology Group.

United Nations Development Programme (UNDP) (1998) *Gender in Development, Mainstreaming Gender*, Geneva: United Nations Development Programme, <http://www.undp.org/gender/capacity/gm_tips.html> (accessed May 2002).

United Nations Development Programme (UNDP) (2001) *Human Development Report 2001*, Geneva: United Nations Development Programme, http://www.undp.org/hdr2001/

Varley, A. (ed.) (1994) *Disasters, Development, Environment*, Chichester: Wiley.

Waddell, E. (1977) 'The hazards of scientism: a review article', *Human Ecology* 5 (1), 69–76.

Watts, M. (1983) 'On the poverty of theory: natural hazards research in context', in K. Hewitt (ed.) *Interpretations of Calamity*, London: Allen and Unwin.

Wilson, J., Phillips, B. and Neal, D. (1998) 'Domestic violence after disaster', in E. Enarson and B.H. Morrow (eds) *The Gendered Terrain of Disaster: Through Women's Eyes*, Westport, Conn.: Praeger.

# 5 Natural disasters, adaptive capacity and development in the twenty-first century

*Mohammed H.I. Dore and David Etkin*

## Introduction

There is widespread agreement that climate change poses a serious threat to the well-being of the earth's environment and the strength of its economies. The build-up of greenhouse gases such as carbon dioxide requires action be taken to reduce the consumption of fossil fuels in industrialized countries if the rate of climate change is to be reduced to a level which will allow the implementation of adaptations that will reduce negative impacts. These countries account for more than two-thirds of annual carbon dioxide emissions world-wide. Developing countries have much lower per capita emissions, and are more concerned with providing for the basic needs of their people, rather than climate change. Because it is projected that by the year 2020, the emissions from developing countries will exceed those of industrialized countries, the time is right to pursue a more sustainable path of development.

Natural disasters occur when an event such as an earthquake or storm reveals social vulnerability, and consequent damage to the physical and social fabric exceeds the ability of the affected community to recover without assistance. Societies respond to a disaster by means of three overlapping activities: response and recovery, mitigation and preparedness. These activities alter future vulnerability, reducing risk if they are done well or not if they are done badly. This relationship is shown in Figure 5.1 (Etkin 1999) and depicts a dynamic, interactive system, composed of both natural and social forces.

None of the boxes in this figure are static. It is the box in the top right corner, labelled 'Hazard', that is relevant to climate change. The frequency and intensity of heat waves, cold waves, droughts, floods, tornadoes, etc., are likely to change in the future, altering our *risk*. Of course, *vulnerability* changes over time as well, due to a host of socio-economic and environmental factors. While there are other factors that affect the vulnerability of a given society, in this chapter we concentrate on the impacts of climate change that are likely to be revealed largely through changes in the pattern, severity and frequency of climate-related natural disasters.

Working Group I of the United Nation's Intergovernmental Panel on Climate Change (IPCC) has concluded that the globally averaged surface

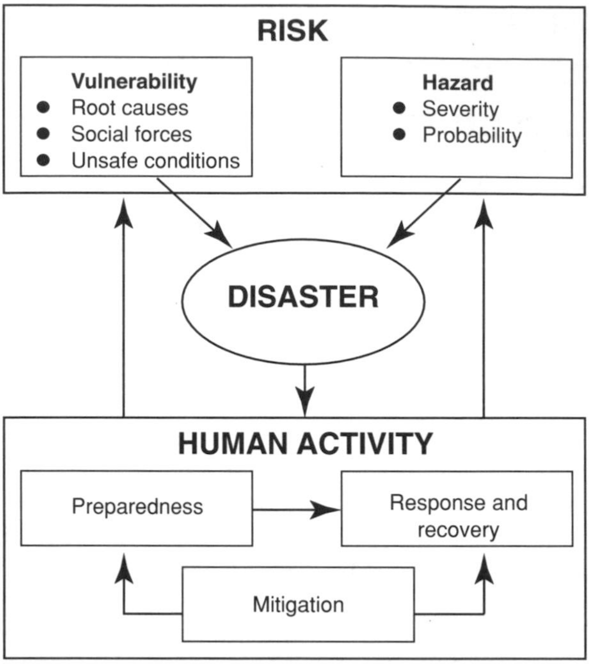


*Figure 5.1* Disaster cycle.
Source: Etkin (1999).

temperature has increased by 0.6 °C (+ or −0.2 °C) since the late nineteenth century (IPCC 2001a). For the range of scenarios developed in the IPCC *Special Report on Emission Scenarios*, the globally averaged surface air temperature is projected to increase from 1.4 to 5.8 °C by 2100, relative to 1990 levels. Average sea levels over the globe are projected by models to rise by 0.09–0.88 m by 2100 (IPCC 2001a). These figures illustrate the uncertainty associated with these global projections. Also, changes will vary regionally in very significant ways. In addition, there will be changes in the variability of climate, and changes in the frequency and intensity of some extreme climate phenomena. If, as scientists now expect, 'positive feedback' occurs in the environment due to changes that have already happened, then warming could progress even faster. People living in the world's poorest regions are likely to be even more at risk than they are now. By 2080 it is expected that over 3 billion people across Africa, the Middle East and the Indian subcontinent will suffer an increase in water stress (IPCC 2001b). Agricultural yields in Africa are expected to drop and hunger is expected to rise. Both droughts and floods are expected to increase in frequency (IPCC 2001b).

The world has seen more natural disasters in 2000 than in previous years of the decade; the number of affected people went up to 256 million, compared with an average from 1991 to 2000 of 211 million people per year (IFRCC 2001). A major cause of the increasing number of people being affected by disasters is the increase in the number of hydrometeorological disasters such as floods, windstorms and droughts. In some disaster-prone areas, recurrent weather-related disasters are sweeping away development gains. This leads one to question the efficacy of development strategies that do not respect the phenomenon of climate change. Clearly, future development policy must incorporate an increase in *adaptive capacity*, of which an important component is the ability to devise development policy that takes into account global climate change and its adverse impacts on developing countries.

In this chapter we take the position that without enhanced adaptive capacity, developmental expenditures in low- and middle-income countries will simply be wasted, as traditional-style economic development projects (geared to earning foreign exchange) tend to reproduce the same vulnerability characteristic of earlier development. The chapter is organized as follows. First, we examine the record of natural disasters in the developing countries and provide *prima facie* evidence that development policy must take hydrometeorological disasters into account. The second section is an exposition of the concept of adaptive capacity to climate change. Third, we use individual country examples from three continents to show that nineteenth-century patterns of development that do not enhance adaptive capacity to hydrometeorological disasters will fail. Fourth, we provide policy lessons to development when the degree of adaptive capacity must vary directly with the vulnerability of that country. It argues that development must be of a very different kind in the twenty-first century for it must take into account two interrelated phenomena: global climate change and land use policy. The chapter ends with a brief summary and conclusion.

## Natural disasters in developing countries

The historical record of natural disasters (OFDA/CRED EM-DAT 2001 International Disaster Database) provides evidence that hydrometeorological disasters (HDs) are:

1 increasing in frequency and variability of occurrence in the developing world;
2 more frequent than other forms of natural disasters that strike the developing world.

Figure 5.2, shows an upward trend over time in the frequency of hydrometeorological disasters striking Africa, Asia and Latin America, a phenomenon that we expect will continue. Increased variability of occurrence is displayed by increase in the 95 per cent confidence intervals for the average number of occurrences per year towards the end of the century. The widening confidence interval is a very economical way of demonstrating the increase in variability of HDs. Figure 5.3, shows the extent to which hydrometeorological disasters occur more frequently relative to geophysical and other forms of natural disasters, such as epidemics, famines, insect infestations and

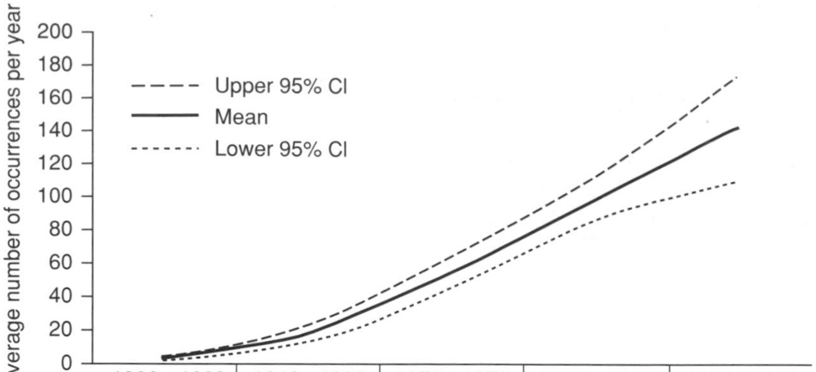


| | 1900 - 1939 | 1940 - 1969 | 1970 - 1979 | 1980 - 1989 | 1990 - 2000 |
|---|---|---|---|---|---|
| Upper 95% CI | 5.15 | 23.41 | 64.39 | 112.11 | 172.65 |
| Lower 95% CI | 2.95 | 13.46 | 45.01 | 87.29 | 111.35 |
| Mean | 4.05 | 18.43 | 54.70 | 99.70 | 142.0 |

*Figure 5.2* Hydrometeorological disasters in developing countries.

Source: OFDA /CRED International Disaster Database (2001).

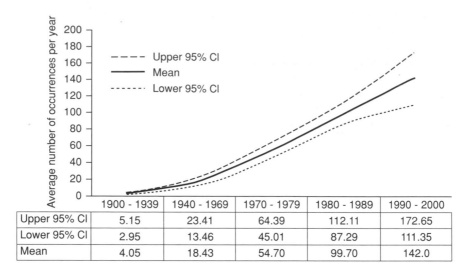

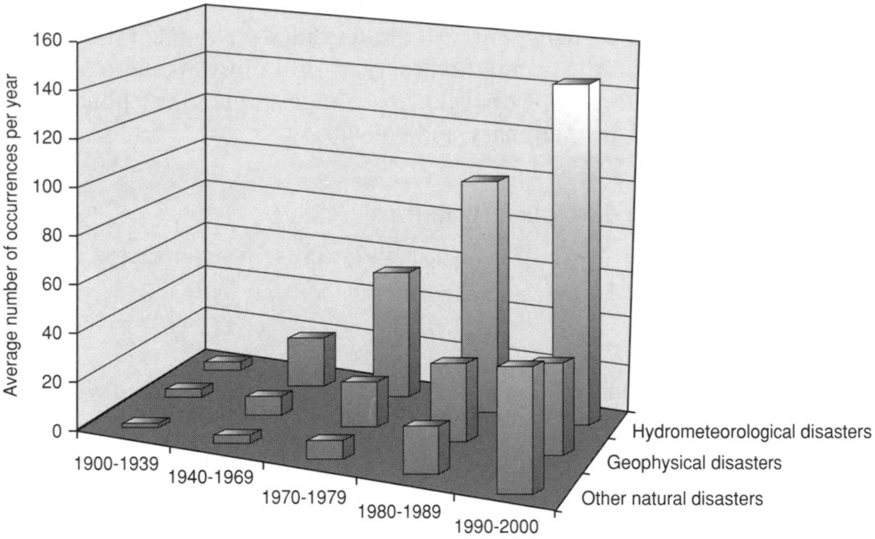


*Figure 5.3* Natural disasters in developing countries, by type.

Source: OFDA/CRED International Disaster Database (2001).

wildfires. In the 1990s an average of 142 weather related disasters (HDs) occurred per year throughout the developing world, compared to an average of 39 geophysical disasters and 51 other natural disasters. Hydrometeorological natural disasters include droughts, extreme temperature events, floods and windstorms. Geophysical natural disasters include earthquakes, slides, volcanic eruptions and wave/surges. Other natural disasters include epidemics, famines, insect infestations and wildfires.

Unfortunately, it is impossible to ascertain from the data in the CRED database whether or not hydrometeorological disasters have become more severe over time, though data from Munich Re suggests that impacts, at least, have indeed been increasing (Munich Re 2001). In the Third Assessment Report by IPCC Working Group 1 (IPCC 2001a) there is an assessment of the confidence in observed changes in extremes of weather and climate during the latter half of the twentieth century and in projected changes during the twenty-first century. The assessment relies on observational and modelling studies, as well as the physical plausibility of future projections across all commonly used scenarios, and is based on expert judgement. The results are provided in Table 5.1. From this information we can conclude

*Table 5.1* Estimates of confidence in observed and projected changes in extreme weather and climate events world-wide

| *Judgemental confidence in observed changes (latter half of the twentieth century)* | *Changes in phenomenon* | *Judgemental confidence in projected changes (during the twenty-first century)* |
|---|---|---|
| 66–90% | Higher maximum temperatures and more hot days over nearly all land areas | 90–99% |
| 90–99% | Higher minimum temperatures, fewer cold days and frost days over nearly all land areas | 90–99% |
| 90–99% | Reduced diurnal temperature range over most land areas | 90–99% |
| 66–90% over many areas | Increase of heat index over land areas | 90–99% over most areas |
| 66–90% over many northern hemisphere mid- to high-latitude areas | More intense precipitation events* | 90–99% over many areas |
| 66–90% in a few areas | Increased summer continental drying and associated risk of drought | 66–90% over most mid-latitude continental areas (lack of consistent projections in other areas) |
| Not observed in the few analyses available | Increase in tropical cyclone peak wind intensity† | 66–90% over some areas |
| Insufficient data for assessment | Increase in tropical cyclone mean and peak precipitation intensities | 66–90% over some areas |

Source: IPCC (2001a).

* For other areas there are either insufficient data or conflicting analysis.

† Past and future changes in tropical cyclone location and frequency are uncertain.

that, although we may not have sufficient information to make judgements about the trends in the severity of hydrometeorological disasters, if climate projections are plausible we will probably see hydrometeorological disasters of increased intensity in the twenty-first century.

For developing nations climate change presents special challenges. Negative effects of climate change are adversely distributed against developing countries (IPCC 1995). In addition, most developing countries do not have adequate human, financial and technical resources to cope with the effects of climate change. Being able to cope requires adaptive capacity. In the next section we explore some conceptual issues in defining adaptive capacity, and some benchmark which we call 'baseline adaptation'. We also consider optimal adaptation, i.e. the level of adaptation that is appropriate for a country, given its resources and vulnerabilities.

## The concept of adaptive capacity

### *Baseline adaptation*

In what follows, all use of the word 'adaptation' refers to baseline adaptation to current (or past) climate, to normal variability of climate, as defined by the WMO. One might ask what are the *necessary and sufficient* conditions for adaptation? When is adaptation complete? To begin, we might ask in what ways are *developed* countries adapted to current climate. First of all, they have the technical know-how to understand climate. Second, they have resources to devote to research in climate and into the consequences of normal climate variability, including extremes. Third, they develop the necessary technology to *cope* with climate (from building codes to building materials that can withstand a particular climate). Fourth, they share risk through government disaster assistance programmes and through the insurance markets, which are willing to assume many of the risks associated with climate, when the actuarial basis for calculating risk premiums exists. Fifth, the insurance markets mediate moral hazard problems through mechanisms such as (a) a deductible minimum, (b) rebates for minimizing damage (e.g. rebate for having a fire extinguisher), (c) premium reductions for no-claims. Sixth, developed countries invest resources (in a variety of ways) in emergency response at all levels of government. Each of these six points need elaboration. When all of the necessary conditions are specified, together they define *adaptive capacity* at an institutional level.

We have tried to move towards a definition of adaptive capacity by reference to developed countries, just as a gap in per capita income is used to differentiate between developed and developing countries. Thus it may be reasonable to assume that the developed countries are in general 'adequately' adapted. In fact, many case studies exist to show that even developed countries are not well adapted to climate extremes at times. However, that is outside the scope of this chapter. For the purpose of this exercise we can assume that developed countries in general are adapted to current climate and therefore the concept of adaptive capacity can be derived by reference to developed countries. For some developing countries adaptive capacity may be minimal if there is evidence of repeated losses due to variability of current climate.

The six points listed above are all part of the *necessary* conditions for adaptation, but they are by no means sufficient. Necessary conditions (by definition) contribute towards successful adaptation. Next we will consider sufficient conditions for baseline adaptation. Clearly there are no developed countries that are so *completely* adapted to climate as to reduce damage due to climate to zero. Nor should there be, as over-adaptation will have an opportunity cost. No matter how well adapted a country may be to its current climate, there can still be damage due to rain, hailstorms, wind, etc. Therefore, there exist no *sufficient* conditions for complete baseline adaptation. Even if every building were built up to the standards of best available technology, there would still be costs of accidents that could be attributed to climate.

We are led to the conclusion that baseline adaptation should be restricted to just necessary conditions. Most developed countries have adapted to some 'current' climate regime, where current refers to some given *norms*, such as constructing buildings and infrastructure for historically given *return periods* for particular types of natural disasters. We can call this level of in-built protection, the *baseline adaptation*. When a country meets the norms, it has adapted to some baseline climate. In addition, as the evidence of losses due to climate change mounts, even the developed countries will have to adapt further, up to some adequate level. It is the necessary conditions for baseline adaptation that will lead us towards a definition of *socially optimal adaptation*. Some adaptations are unnecessary: for example, we do not all need to live in buildings built like nuclear bunkers. Some adaptation measures cannot be justified in terms of their cost in relation to the risk. A country could also be under-adapted if there are adaptation measures available which, if used, could yield net benefits in the form of reduction in losses due to damages. The set of necessary conditions together defines adaptive capacity. We can expect adaptive capacity to be vector valued and hence not simple; that is, there are at least six components of the vector that together define adaptive capacity. It would be simple only if we had the complete set of necessary and sufficient conditions to characterize adaptive capacity. The relationship between adaptation capacity and socially optimal adaptation is explored on pp. 82–83.

We can measure adaptive capacity by developing an ordinal index of the six necessary conditions. This index would be used as an ordinal scale on each necessary condition. For example, country A invests more in climate research than country B; country C is more open to insurance markets than country D. Thus one can envisage such pair-wise ordinal comparisons among developing countries. Of course, like the Human Development Index this ordinal scale can be converted into a cardinal index if the relative weights of the six conditions are given exogenously to reflect the importance of each for the practice of adaptation. Alternatively, we can choose the same index and normalize it by treating baseline adaptation in Annex 1[1] countries to be 100. (The World Bank does something similar in calculating purchasing-power parity of world currencies.) International comparisons can then be made and all non-Annex 1 countries can then be given some ranking by computing the damage caused by climate in the past five years as a ratio of GDP. Damage costs must be estimated as economic and monetary losses, plus imputed costs of life and health (mortality and morbidity) and imputed environmental damage costs attributable to climate (Dore and Etkin 2000). The country that has the largest damage costs relative to its GDP will then have the lowest rank in terms of baseline adaptation.

### *Socially optimal adaptation*

What is the relationship between adaptive capacity and socially optimal adaptation? What are the dimensions of socially optimal adaptation (SOA)? Is SOA time-and-place dependent? Consider some examples first. What is SOA for one country will not be the same for any other country. For example, the Philippines is the most disaster-prone country in the world. The SOA for the Philippines will depend on its adaptive capacity. As its adaptive capacity changes (increases, we hope), so does its SOA. Therefore SOA will not be static, as it changes over time. It also changes with geographical location, as the set of necessary conditions will vary from one location to another. We can now differentiate between adaptive capacity and SOA. The SOA is the 'prudent' and feasible level of expenditure that a country might want to carry out, *given its baseline adaptive capacity at a particular time*. What is prudent will depend on the values of a society. Thus a 'no regrets policy' suggests a higher level of adaptation.

Clearly the concepts of baseline adaptive capacity and the socially optimal adaptation are related. On the one hand, the World Bank and other development assistance agencies would want to see the adaptive capacity of a developing country increase. On the other, the socially optimal level of adaptation at a particular time would depend on local level research to determine concrete adaptation measures for a particular country. Therefore SOA can be determined only after on-site research, taking into account a number of constraints to expenditures to adaptation.

It might be useful to note that as we are considering *current climate*, SOA does not involve risk analysis, as normal climate risk can be 'decentralized' to the market mechanism through insurance markets. However, not all risks are insured; for example, in Canada there is no flood insurance. In some countries there is no well-developed insurance market, such as Japan – the government covers most risks. Many developing countries do not have a culture of insurance, e.g. Mexico. Even in Canada and the US the government assumes many risks. But SOA will *depend* on the extent to which the insurance market has assumed the particular climate risk. Therefore it is only when there have been barriers to international capital mobility that the insurance market will not be a participant in normal climate risk insurance. The only real social problem arises when the global insurance market is oligopolistic and participates in a price-discriminatory way in a particular developing country. A small developing country cannot enforce 'competitive' behaviour on the part of the global insurance industry. In that case, there may be a role for international institutions, such as the World Bank, to step in and help by subsidizing insurance. In this, the World Bank would be helping to correct market distortions.

In the above, we were confined to macro-level baseline adaptation. But as vulnerabilities are almost always sectoral, and that is where damage due to *lack* of adaptation occurs, it would make sense to extend the above analysis to the sectoral and project levels.

### *Baseline adaptation at the sector and project levels*

The lack of adaptation shows up, of course, in damage costs and loss of human life. That is, the level of adaptation is below the socially optimal level. Sub-optimal adaptation in developing countries can be seen in the following sectors:

1 Public infrastructure: public utilities such as electricity supply, water supply and wastewater management, ports and harbours (if any), roads and bridges, telecommunications, hospitals, schools and government administrative buildings.
2 Agriculture, forestry and watersheds.
3 Tourism, national parks and the environment.

#### *Public infrastructure*

For developing countries (and also for other countries), an adequate level of infrastructure is important for development. As much of this infrastructure is of a public nature (see point 1 in the list) it might be legitimate to view that infrastructure to be a priority for adaptation to climate (and even to climate *change*). It can also be shown that much of it is publicly provided and is necessary for the private sector to thrive. How well this sector is adapted to current climate can be estimated as follows: (1) take the operating budgets (at all levels) for 'normal wear and tear' expenditures and compute a five-year moving average; (2) if in any one year a particular expenditure is above the moving average, determine whether or not the additional expenditure was climate induced. If climate-induced expenditure is low, then the sector is 'well adapted'.

#### *Agriculture and forestry*

The income from these sectors will be subject to price fluctuations as well as to changes in the physical volume of output. Once price fluctuations are controlled it is possible to determine fluctuations in physical volume of output caused by rainfall variability, temperature-induced insect damage and soil erosion due to wind and rain. One must take care not to include losses in output due to poor land use practices that are not climate related. Countries would be well adapted if climate-induced changes are small or mitigated by switching crops due to intelligent use of climate forecasts. Of course, in some countries there may be *no* expenditures to offset the adverse effects of climate. But there may be evidence of famines, mass starvation, and inflows of food aid. All that evidence indicates poor adaptation to current climate.

### *Tourism*

Many countries in the Caribbean rely on their climate for the flow of tourists during the northern hemisphere's winter. Their tourist industry is thus well adapted to climate. However, in past El Niño years most Caribbean countries suffered losses in tourist revenue. Adaptability would require them to offer other attractions to tourists during cold El Niño years, when the Caribbean beaches are not popular. Casinos and fishing tournaments, and other substitutes in such years, could cushion revenues. A country that has a 'cushion' policy would have a well-adapted tourist sector.

### *Project level adaptation*

The greatest benefit to developmental policy would occur when the above sectors are disaggregated into projects and the adaptive capacity of each project is considered to determine whether or not adaptation is socially optimal. At the project level, one might consider, for example, flood control structures, or irrigation measures in a drought area. For a new developmental project it would be advisable to consider not only adaptation to current climate, but also to climate *change*. In the example of flood control structures it would be prudent to ask how the flood control structure should be modified to make it resilient to the additional stresses that could be attributed to climate change.

## ***Baseline adaptation and adaptation to climate change***

Information on baseline adaptation would be a good adjunct for donor agencies who are considering funding for development projects but do not wish to expose these projects to the additional stresses of expected climate change. Indeed, once the baseline adaptation is considered, it might be possible for both the World Bank and Global Environment Facility (GEF) to co-operate in determining the marginal contribution that GEF might need to make in order that the risk factors associated with climate change do not jeopardize new development projects.

To sum up, in this section we have argued that no country is completely adapted to its present climate so as to reduce damage due to climate to zero.

Also, we distinguished two subsets of adaptive capacity – namely, baseline adaptive capacity and socially optimal adaptation to climate change. In addition, we have stated six necessary conditions that reflect adaptive capacity. But we concluded that there is no way of stipulating the necessary and sufficient conditions for adequate or optimal adaptation to climate change. Baseline adaptive capacity reflects a country's ability to contend with some normal range of climate variability that is typically incorporated in that country's building codes and best practices in construction of buildings and other social infrastructure. Although baseline adaptation is multivalued, it would be possible to construct a cardinal index, given appropriate weights.

On the other hand, socially optimal adaptation is a dynamic concept that would depend not only on geopolitical and economic factors, but also on a country's existing baseline adaptive capacity. Socially optimal adaptation would then change over time, as new information about climate-induced damage becomes available and as known technology changes.

## Case studies

In this section we document the experiences of three developing countries that lack adaptive capacity and suffer periodic losses from hydrometeorological disasters. The three countries are drawn from Asia, Latin America and Africa. The concrete examination of these case studies brings the theoretical discussion closer to policy purposes.

### *Bangladesh*

Climate change will impose significant stress on resources throughout the Asian region. Asia has more than 60 per cent of the world's population; natural resources are already under stress and the resilience of most sectors in Asia to climate change is poor. Consider the example of Bangladesh: it has one of the ten highest population densities in the world and lies at the combined delta of three major rivers. Most of the country is at or near sea level. Each year floods and mudslides from monsoon rains and cyclones displace millions of people. In August of 2000 Bangladesh was struck by flooding as a result of the worst monsoon rains in 20 years. The flooding affected some 2.5 million people, according to the International Federation of the Red Cross (IFRC), and resulted in an estimated US$500 million in economic losses according to the US government's National Oceanic and Atmospheric Administration (NOAA). NOAA reported that 200 villages were submerged, and widespread damage to crops occurred. The IFRC reported that 731,459 houses were affected; that 385,398 acres were flooded; and that 31 people lost their lives (EM-DAT 2001).

Despite domestic and international efforts, Bangladesh remains one of the world's poorest, most densely populated, and least developed nations. Although more than half of GDP is generated through the service sector, nearly two-thirds of Bangladeshis are employed in the agriculture sector, with rice as the single most important product. Major impediments to growth, besides frequent cyclones and floods, are inefficient state-owned enterprises, inadequate port facilities, a rapidly growing labour force that cannot be absorbed by agriculture, delays in exploiting energy resources (natural gas), and insufficient power supplies.

Adaptive capacity of human systems is low and vulnerability is high in Bangladesh as well as the other developing countries of Asia. Most Asian countries relied on economic development prescriptions that took no account of climate change. Twenty-first century development strategies that respect the phenomenon of global climate change will have to take the features listed in Box 5.1 into account.

### *Venezuela*

According to the IPCC there is ample evidence of climate variability at a wide range of timescales all over Latin America, from intra-seasonal to long-term (IPCC 2001a). In many sub-regions of Latin America this variability

**Box 5.1 Projected climate change impacts for Asia**

- Extreme events have increased in temperate and tropical Asia, including floods, droughts and tropical cyclones.
- Decreases in agricultural productivity and aquaculture due to thermal and water stress, sea-level rise, floods and droughts and tropical cyclones would diminish food security in many countries to the arid tropical and temperate Asia; agriculture would expand and increase in productivity in northern areas.
- Runoff and water availability may decrease in arid and semi-arid Asia but increase in northern Asia.
- Sea-level rise increases in the intensity of tropical cyclones in low-lying coastal areas of temperate and tropical Asia; increased intensity of rainfall would increase flood risks in temperate and tropical Asia.
- Climate change would increase energy demand, decrease tourist attraction and influence transportation in some regions of Asia.
- Poleward movement of the southern boundary of the permafrost zones of Asia would result in a change of thermokarst and thermal erosion with negative impacts on social infrastructure and industries.

Source: Adapted from *Climate Change 2001: Impacts, Adaptation and Vulnerability* (IPCC 2001b).

in climate is normally associated with phenomena that already produce impacts with important socio-economic and environmental consequences that could be exacerbated by global warming and its associated weather and climate changes. For example, on 15 and 16 December 1999 the equivalent of two years of rain hit the mountainous coastline of Venezuela. In Vargas State, landslides destroyed 5,500 homes, damaged 25,000 others and wrecked infrastructure. Between 80,000 and 100,000 people were affected, and up to 30,000 died. The United Nations estimated economic damage at US$1.9 billion (EM-DAT 2001). Due to the global weather phenomenon of La Niña, the high-pressure front which usually sits over Venezuela late each year moved north. As a result, fronts with low water-filled clouds moved across Venezuela's northern coast. When they met the 2,700 metre-high Avila mountain range that lies between Vargas State and the capital Caracas, torrential rain was the result. Floods and mudslides hit eight states, but Vargas suffered 80 per cent of the damage and 99 per cent of the death toll. In Caracas, mudslides and floods damaged highways and informal settlements on slopes, killing 59, destroying 1,842 homes and damaging a further 2,261 according to Venezuela's Civil Defence. In the eastern part of Miranda State El Guapo reservoir burst, releasing an 18-metre-high wall of water. The resulting wave carved a mile-wide path of destruction on its way to the ocean, flooding 30,000 hectares and demolishing 790 homes, 60 per cent of the crops, the main water system and a key bridge. Of the estimated US$1.9 billion in damage, 30 per cent was in infrastructure – especially in water and sanitation, transport and communications (IFRCC 2001). The petroleum sector dominates the Venezuelan economy, accounting for roughly a third of GDP, around 80 per cent of export earnings, and more than half of government operating revenues. Unfortunately, a weak non-oil sector and capital flight undercut the economy, which is still struggling to rebuild after the floods and mudslides described above.

The adaptive capacity of human systems in Latin America is low, particularly with respect to extreme climate events, and vulnerability is high. Like Asia, much of development policy of Venezuela and the rest of Latin America has been based on old-style development prescriptions. Twenty-first century development strategies that respect the phenomenon of global climate change will have to take the features listed in Box 5.2 into account.

### *Somalia*

Africa is highly vulnerable to climate change. Impacts of particular concern to Africa are related to water resources, food production, human health, desertification and coastal zones, especially in relation to extreme events. A joint impact of land use and climate change will exacerbate desertification. The African nation of Somalia presents an interesting case. Principally desert, from December to February Somalia is subjected to the north-east monsoon that brings moderate temperatures in the north and very hot temperatures

**Box 5.2 Projected climate change impacts for Latin America**

- Loss and retreat of glaciers would adversely impact runoff and water supply in areas where glacier melt is an important water source.
- Floods and droughts would become more frequent, with floods increasing sediment loads and degrading water quality in some areas.
- Increases in the intensity of tropical cyclones would alter the risks to life, property and ecosystems from heavy rain, flooding, storm surges and wind damages.
- Yields of important crops are projected to decrease in many locations in Latin America, even when the effects of $CO_2$ are taken into account; subsistence farming in some regions of Latin America could be threatened.
- Coastal human settlements' productive activities, infrastructure and mangrove ecosystems would be negatively affected by sea-level rise.
- The rate of biodiversity loss would increase.

Source: Adapted from *Climate Change 2001: Impacts, Adaptation and Vulnerability* (IPCC 2001b).

in the south. From May to October Somalia is under the influence of the south-western monsoon, leading to torrid conditions in the north and hot conditions in the south. Somalia experiences irregular rainfall and is hot and humid between monsoons. Thus the country is susceptible to both droughts and flooding. In January of 2000 Somalia was struck by drought that, according to NOAA, left 21 dead and affected 1.2 million people; the IFRC reported widespread food insecurity as a result of the drought. In November of the same year the country was subjected to flooding which, according to NOAA, affected 60 villages and 150,000 people (EM-DAT 2001).

Somalia is one of the world's poorest and least developed countries, with few resources. Agriculture is the most important sector, with nomadic and semi-nomadic livestock farming accounting for about 40 per cent of GDP and about 65 per cent of export earnings. Livestock and bananas are the principal exports; sugar, sorghum, corn, fish, and qat are produced for the domestic market. There is a small industrial sector, based on the processing of agricultural products, which accounts for 10 per cent of GDP, but most facilities have been shut down because of civil strife. Moreover, ongoing civil disturbances in Mogadishu and outlying areas have interfered with any substantial economic advance, even with international aid arrangements.

Adaptive capacity of human systems in Africa is low due to lack of economic resources and technology, and vulnerability is high as a result of

**Box 5.3 Projected climate change impacts for Africa**

- Grain yields are projected to decrease for many scenarios, diminishing food security especially in small food-importing countries.
- Major rivers in Africa are sensitive to climate variation; average runoff and water availability would decrease in the Mediterranean and southern countries of Africa.
- Desertification would be exacerbated by reductions in average annual rainfall, runoff and soil moisture, especially in southern, North and West Africa.
- Increases in droughts, floods and other extreme events would add to stresses on water resources, food security, human health and infrastructure and would constrain development in Africa.
- Significant extinction of plant and animal species is projected and would impact rural livelihoods, tourism and genetic resources.
- Coastal settlements (Gulf of Guinea, Senegal, Gambia, Egypt, and along the east–southern African coast would be adversely impacted by sea-level rise through inundation and coastal erosion.

Source: Adapted from *Climate Change 2001: Impacts, Adaptation and Vulnerability* (IPCC 2001b).

heavy reliance on rain-fed agriculture, frequent droughts and floods and poverty. Twenty-first century development strategies that respect the phenomenon of global climate change will have to take the features stated in Box 5.3 into account.

## Policy lessons

Recommendations for economic development and the traditional advice from the World Bank and the IMF have tended to focus on stimulating local markets, production of exports for foreign exchange and 'structural adjustment' to reduce the government's external debt and trim the size of the government bureaucracy. There is concern among some aid agencies that the work of the World Bank and the IMF has been counterproductive in the face of disasters and has not helped to increase adaptive capacity, but has rather contributed to an ongoing disaster cycle. Much of these traditional prescriptions will fail in the twenty-first century if development policy does not also include the enhancement of adaptive capacity to climate change. An enhanced adaptive capacity will also help protect previous capital accumulation and improve the effectiveness of new development efforts. Many developing nations are likely to be devastated by an increased frequency of hydrometeorological disasters, and aid agencies will see the same hands begging for help, *more often*. The fact that some of the increase in disasters is due to climate change is likely to be forgotten. It is therefore imperative that attention be directed specifically to increasing adaptive capacity.

## Summary and conclusion

The adverse effects of climate change will be evident in the hydrometeorological disasters that will increase in variability and severity world-wide. In order to lessen the impacts of climate change in developing nations, development must account for an increase in adaptive capacity. Consequently, development policy needs to be updated by incorporating climate change implications in development proposals. Baseline adaptive capacity and socially optimal adaptation are related, but the latter depends on research done at the local level to determine the appropriate measures for a particular country.

No country is so completely adapted to its climate as to reduce damage to zero. Projects should be evaluated on the basis of adaptation to current climate and also to climate change. As a result of climate change, one of the most important aspects of development policy is that it be directed specifically to increasing adaptive capacity.

## Note

1 Annex 1 countries are fully industrialized nations and countries that are undergoing the process of transition to a market economy.

## References


Dore, M.H.I. and Etkin, D. (2000) 'The importance of measuring the social costs of natural disasters as a time of climate change', *Australian Journal of Emergency Management* 15 (3), 46–51.

EM-DAT (2001) The OFDA/CRED International Disaster Database, Universite Catholique de Louvain, Brussels, Belgium (available at www.cred.be/emdat).

Etkin, D. (1999) 'Risk transference and related trends: driving forces towards more mega-disasters', *Environmental Hazards* 1, 69–75.

Intergovernmental Panel on Climate Change (IPCC) (1995) *Impacts, Adaptations, and Mitigation of Climate Change*, Contribution of Working Group II to the Second Assessment Report of the Intergovernmental Panel on Climate Change, Cambridge: Cambridge University Press.

Intergovernmental Panel on Climate Change (IPCC) (2001a) *Climate Change 2001, The Scientific Basis, Summary for Policy Makers*, Contribution of Working Group I to the Third Assessment Report of the Intergovernmental Panel on Climate Change, Cambridge: Cambridge University Press.

Intergovernmental Panel on Climate Change (IPCC) (2001b) *Climate Change 2001: Impacts, Adaptation and Vulnerability*, Contribution of Working Group II to the

Third Assessment Report of the Intergovernmental Panel on Climate Change, Cambridge: Cambridge University Press.

International Federation of Red Cross and Red Crescent Societies (IFRCC) (2001) *World Disasters Report 2001*, Bloomfield Conn.: Kumerian Press Inc.

Munich Re Group (R&D/Geo) (2001) *Topics 2000 Annual Review: Natural Catastrophes 2000*, Munich: Munich Re Group.

# Part III

# International exchange and vulnerability

# 6 Actors in risk

*Ian Christoplos*

## Introduction

In the past, disaster management was strongly infused with implicit assumptions that there were clear-cut 'normal' roles for the state, the private sector and civil society. Disaster mitigation and preparedness (DMP) mainly consisted of technical and material inputs to prepare for the predetermined roles of these different actors. We felt that we at least knew who *should* do what when dealing with disaster mitigation, preparedness and response. Methods to deal with natural disasters were debated, and quality was a topic for concern, but the goal posts were not in question. In recent years, however, changes in the nature of disasters, in the structure of international aid architecture and in the discourse on humanitarian response, have compelled us to reflect critically on who should be doing what before, during and after a disaster strikes. The implications of globalization and structural adjustment have introduced a growing ambiguity into what had been seen as self-evident set-piece roles for states, non-governmental organizations (NGOs), and the private sector.

Behind the discourse on disasters, neo-liberal policies have taken hold and pressured states to assume a narrower set of responsibilities. Particularly with respect to DMP, it has become increasingly clear that this has been done without making clear who will shoulder these duties instead. Critics of globalization are responding to failure to prepare for and mitigate disasters with calls for stronger state institutions and greater control by 'communities', but these critics are also vague about how this is to be accomplished in the face of frightening demographic scenarios, spiralling conflicts and a steady decline of resources for public goods, such as DMP. Decentralization has placed greater responsibilities on local institutions to deal with disasters, but performance has rarely lived up to rhetorical aims of bringing power to the grassroots. Local politicians are, if anything, even more hostage than their colleagues at higher levels to demands to deal with immediate problems rather than distant threats. Ambiguity is perhaps greatest in the marginal areas most affected by natural disasters, where memories of the days of effective public authority are fading. In places like these, government capacity is disappearing and neither the NGOs nor the private sector are filling the gap.

Gaps exist because DMP is easy to ignore most of the time. When the threat of a disaster is not pressing, politicians and the general public are more concerned with everyday life. When a disaster does occur, those responsible for response (at least in the South) may receive a sudden influx of money, but they too often lack resources and support to raise the issue of dealing with hazards after the media attention has subsided and both collaborating institutions and flows of funding have returned to normal. Risk lacks a clear home in development and aid architecture. It is not a 'sector' with self-evident champions to fight for funding, and it is therefore easily pushed aside as being somebody else's business.

Perhaps the greatest gap is the grey area between development actors and humanitarian agencies. Development actors tend to downplay vulnerability due to traditional emphasis on economic development. Poverty is assumed to be about lack of income, not about addressing or avoiding acute human suffering. Disasters should be acknowledged as indications that the development experts are missing something. Instead, they are usually glossed over as troublesome minor glitches in economic models. Humanitarians have for some years been under pressure to address risk in relief operations through so-called 'developmental relief', but short-term funding cycles and the complex interrelations between risk and broader development issues discourage concerted and consistent efforts within bureaucracies that are oriented toward filling temporary service provision gaps. Relief administrators simply lack the skills to confront questions of how to avoid the next disaster, and are loath to enter the political fray where genuine risk management must be anchored.

In order to face increasing societal hazards, risk must be perceived as part of both development and humanitarianism. It cannot remain 'somebody else's business' that regular development and humanitarian policy-makers and practitioners needn't worry about. It must become part of 'their' ongoing discourses. There are three different discourses into which risk must find new meaning if it is to become the business of more than a small group of frustrated scientists and individuals scattered through the humanitarian and development trades. In all three of these discourses there are some positive signals.

- First, the humanitarian discourse on changing roles and alternatives to the relief–development continuum has generally acknowledged that disasters cannot be assumed to be small bumps on the grand road to development. New ways of dealing with chronic and recurrent crises are needed. Risk, therefore, is not something to be left in an easily forgotten 'pre-disaster phase'.
- Second, the rising economic cost of disasters due to hazards related to long-term global phenomena, such as climate change and increased population, have raised awareness among development practitioners that

the risks facing development efforts must be addressed if development itself is to proceed as planned. The costs are being acknowledged as simply too large to ignore.

- Third, efforts to understand how poor people perceive and seek to deal with their poverty have highlighted how their livelihood strategies are often more about addressing vulnerability and handling shocks than about 'escaping' from poverty *per se*. If 'we' are going to synchronize our work with 'them' (the poor), risk must be put at the top of the agenda.

DMP is thus a goal, rather than a particular type of project. It should be addressed as a continuous and integrated process involving a broad spectrum of actors. Opportunities for improving mitigation and preparedness can be found in all sectors of activities, including relief, rehabilitation, development, health, agriculture, social protection and community services. DMP is the *result* of a wide range of activities and resources. DMP may lose relevance if it is tucked away in a project. Nonetheless, aid structures tend to force agencies to projectize. Rather than creating conditions for a broad-based 'culture of prevention', the architecture of aid creates pressures to design work within project structures that single out limited sets of activities within a specific time-frame. As a result, activities are rarely situated within broader strategic thinking about why people are vulnerable. The challenge is to find ways that take advantage of the opportunities which DMP projects provide, without getting pulled into the contract culture of designing DMP efforts around donor objectives and funding structures.

## Actors and action

As mentioned above, the humanitarian, economic development and livelihood discourses of relevance to DMP have tended to be situated among several different sets of actors. Synergy will depend on expanded understanding of the conceptual frameworks through which risk management is perceived among these institutional actors, and how these perceptions are manifested in operational priorities and programmes. It is essential to understand *what* these actors are already doing, and *why*, before prescribing solutions. DMP has been particularly susceptible to dramatic calls for 'somebody' to solve the problem fast. This is a tendency that is greatly enhanced by media attention for a short period directly after a disaster. Unfortunately, the automatic reaction is usually to press for technocratic package solutions based on false assumptions about the capacities and goals of different actors. Grand projects are launched without having been anchored in existing institutions.

An actor-oriented perspective (see e.g. Long and Long 1992) involves looking at how these different actors make sense of the crises they are confronted with (Christoplos 1998). World Bank experts, NGO volunteers, village mayors and refugees are all struggling to come to grips with globalization based on their own personal and institutional norms. Disasters, as evidence of far-reaching systemic failure, create moral dilemmas as the weaknesses in different development models are laid bare. Aid bureaucrats and farmers realize that something is wrong. Disasters, therefore, can be a stimulus for a fresh perspective on the implications of each individual's own institutional position on how globalization impacts on human suffering. For this reason it is inappropriate to jump to conclusions about what the 'World Bank' or 'NGOs' think about disasters and development. They do not think. The people working within these different types of organizations have different entry points into the debate on risk and globalization. They have different vested interests and different cultural perspectives, but their ultimate conclusions about how to confront risk are far from fixed.

### *Non-governmental organizations*

The first of the actors are the non-governmental organizations (NGOs), which work in both humanitarianism and development. NGOs play a major role in keeping acute human suffering on the global agenda. It is in their humanitarian and development actions that NGOs are most visible to the general public. They are also increasingly engaged in advocacy. This advocacy focus is growing due to several trends:

- NGO funding, although large, is fragmented into small and often short-term service provision projects that in many cases have limited impact on the wide-ranging structural challenges of risk reduction (Rocha and Christoplos 2001).
- There is a growing concern among some agencies that they are unintentionally contributing to the neo-liberal project of weakening state responsibilities by taking over the operational duties of the state in humanitarian emergencies, and even afterwards. Advocacy is a way of avoiding and hopefully counteracting such a danger.
- NGOs have realized that their growing contractor relationship to donors has begun to compromise their own core values as watchdogs of public authority (Hulme and Edwards 1997; Fowler 2000).

When disasters strike, NGOs are therefore usually the first to point out the need for better disaster preparedness. But do they practice what they preach in their own operational priorities? The involvement of NGOs in risk management varies according to internal factors, such as policy frameworks and the linkages between relief and development departments, and external factors, such as donor priorities. Benson *et al.* (2001) point out that NGO staff themselves have limited understanding of basic DMP concepts. Even those who would prefer to promote and integrate risk reduction, lack tools to assess, monitor and evaluate DMP initiatives, which are often perceived as being too vague to promote in planning discussions.

NGOs frequently have stated principles that support efforts to make the necessary linkages, but they are hostage to the same structural obstacles as the aid donors with which they work (Fowler 2000). They too raise their funds either through donors' relief or development departments, or by appealing to traditional constituencies in the general public. Here, NGOs are in many respects dependent on their symbiotic relationship with the media, where simple messages about dying babies are more effective than statistics about vulnerability. Largely because of this, there is no popular movement in the northern countries where NGOs raise their funds dedicated to reducing risk. Even the Red Cross has a far easier time raising funds for traditional relief than for mitigation or preparedness. It is not so difficult to bring up such questions in the countries most affected by disasters (IFRC 2002), but sister Red Cross societies usually do not give disaster preparedness top billing in media campaigns.

The ways in which NGOs remain tied to media and donor priorities, and fail to live up to their rhetorical aims, ultimately stem from their structural integration into aid bureaucracies. Despite claims of having a 'partnership' with local civil society, there are few mechanisms to ensure that northern agencies are accountable to the local civil society organizations with which they work, much less to disaster victims (Fowler 2000; Callamard and van Brabant 2002).

It must be stressed that some humanitarian NGOs actually give DMP low priority as a matter of principle. Over 160 NGOs have formally accepted a Code of Conduct that begins by stating that 'The humanitarian imperative comes first' (SPHERE 2000). Their task is to save lives and address acute human suffering. DMP is frequently placed low on the agenda since a shift in resources towards a hypothetical long-term objective is perceived as conflicting with their ethical commitments to immediate life-saving response. This focus on core values is further reinforced where humanitarian neutrality demands a distance from the political sphere. The other chapters in this volume point out that effective DMP is reliant on an effective state, and that neo-liberal globalization has in many countries devastated the ability of the state to fulfil its duties to ensure the safety of its citizens. Humanitarian agencies are acutely aware of this, since they are involved in cleaning up the aftermath of failed states. At the same time, however, they must be cautious about entering into projects that rebuild the authority of government offices that are occupied (or are still under bombardment) by warlords. Within the humanitarian community, there are growing calls for a return to basic humanitarian principles. For many, that means staying away from the political arena (see e.g. Disasters 2001; Slim 2002). A commitment to better DMP is inevitably a political stance, which is in some significant respects at odds with humanitarian neutrality.

A complicating factor, that few NGOs have confronted in a clear and consistent manner, is the question of whether natural disasters are really humanitarian emergencies. It is often pointed out that natural disasters are different from so-called 'complex political emergencies' (Duffield 1994), but surprisingly little attention is given to defining what they are instead (Christoplos 2000). Agencies frequently have one set of policies and codes for development and another for complex political emergencies, but do not make clear their role in natural disasters where the humanitarian imperative of responding to acute human suffering is apparent but where relations to the state are more ambiguous. In some agencies (Medicins sans Frontiers, for example) there is an internal debate about whether they should respond to natural disasters at all. An even more glaring conceptual gap is that of policies for dealing with natural disasters that occur in the midst of conflict (e.g. drought and earthquakes in Afghanistan during the Taliban regime or the volcanic eruptions in Goma). Is relief enough? If not, who is the duty bearer that the NGO should work with to reduce risk? This is a grey area for both researchers and practitioners.

### *Multilateral and bilateral development institutions*

The second set of actors are the multilateral and bilateral development institutions. Interest in disaster mitigation and preparedness is growing in these organizations (albeit from a very modest level). Infrastructure investment is being related to risk as resilience to 'livelihood shocks' is increasingly acknowledged as an integral part of poverty reduction (Carney 1998; Twigg 2001a).

A new realization of disasters being part of development has been most notable among multilateral institutions. The World Bank has listed 'security', in the sense of reduced vulnerability to shocks and resilience from their effects, as one of the three pillars of poverty alleviation (World Bank 2000). With the establishment of The Disaster Management Facility (DMF) in 1998, the World Bank has aimed at mainstreaming disaster prevention and mitigation practices into all development activities. Large multilateral loans are being granted for infrastructure to control flooding and to protect urban areas. Calls are being made to institute frameworks for creating new development co-operation structures specifically for adapting human settlement and economic activity to climate change (van Aalst and Burton 2000). These new emerging priorities are apparently driven by a realization that urbanization and other globalization-related trends are leading to a concentration of investment in specific geographical areas, and that these investments require new forms of protection. In addition, the NGO-led backlash against globalization has placed a measure of blame on the World Bank and other multilateral institutions for investments that increase environmental risks, such as dams and high-external input agriculture. In order to regain credibility, multilateral institutions have struggled to show that they (and their colleagues in the private sector) can address these risks.

Particularly after Hurricane Mitch, the Inter-American Development Bank (IDB) has also become a major actor supporting mitigation and preparedness programmes. Building on the International Decade for Natural Disaster Reduction (IDNDR) UNDP has launched initiatives to encourage a 'culture of prevention' based on a recognition of vulnerability reduction as a part of the development process. UNDP has included disaster reduction as an integral component of their overall planning framework, and risk and vulnerability is mainstreamed into development and post-disaster recovery through their Disaster Reduction and Recovery Programme (UNDP 1999).

### *The scientific community*

The role of the scientific community in DMP is undergoing a reassessment. The IDNDR was in many respects envisaged as an event wherein scientific knowledge would simply be 'put into practice' to prevent disasters. A failure to achieve a linear relationship between knowing and doing has led to a realization that acting on the advice of research is a complex process where socio-political processes are central (Blench and Marriage 1998). The complexity of the relationship between science and policy formation has been increasingly raised in the development discourse (see Stone *et al.* 2001), but a technocratic faith in the linear applicability of scientific advice in disaster mitigation has proven surprisingly resilient. Some place the blame for the gap between science and practice at the doorstep of the scientific community, who have failed to understand the complex social context of disasters (Hewitt 1996; Bankoff 2001). Others point to failures in the political system (Wisner 2001). The successor to the IDNDR, the International Strategy for Disaster Reduction (ISDR), shows signs of a shift to a more midway position between scientific knowledge and policy formation, and highlights better the roles of vulnerable communities in risk management. The coming years will show whether this new approach can serve to better mainstream the findings of the scientific community in both the ongoing efforts of vulnerable communities themselves and the actions of political actors in policy formation and implementation.

### *The private sector*

The private sector has in recent years been identified as a significant actor in dealing with risk. It has been suggested that the insurance industry may provide a viable channel of resources for both dealing with the impacts of disasters and for promoting risk mitigation through the power of the market (see Salt, Chapter 8 this volume). Whereas in the past insurance was never mentioned in humanitarian and development circles, a realization is emerging that uncertain and limited aid flows will never cover more than a small fraction of DMP needs (Charvériat 2000). Alternative approaches for acting on this realization are being pursued in various World Bank efforts, such as its Market Incentives for Mitigation Investments project, new loan guarantee schemes (Partial Risk Guarantees), exploratory weather insurance mechanisms, and in various studies and publications (Skees *et al.* 2002; Varangis 2001; Kreimer *et al.* 1999; Kreimer and Arnold 2000).

Interest in the role of insurance has expanded (Kreimer and Arnold 2000) as a way to encourage risk reduction through market mechanisms and to transfer risk from individuals and governments to insurance companies and capital markets, thereby alleviating extended hardship after a disaster and disruption to development programmes due to unforeseen expenditure on rehabilitation. Problems of covariate risk, the lack of a tradition of purchasing insurance, and, most importantly, the lack of information by which insurers and households can accurately judge risk present major challenges to the expansion of such mechanisms for poor people in the South.

There have been some expectations that the private sector may even play a significant altruistic role in disaster management as part of hopes that large companies may adopt DMP as a potentially high profile demonstration of their corporate social responsibility. So far, however, there is little evidence that the private sector is taking on such a role on a significant scale. Twigg (2001b) attributes this to the mismatch between the private sector's primary aim of pursuing profits and the altruistic goals of risk reduction. He also draws attention to their lack of willingness to address the underlying social and structural causes of increasing vulnerability, many of which relate to globalization and the role of private capital itself.

### *Local institutions*

The massive but usually underestimated role of local institutions in risk management is starting to be acknowledged. 'Risk averse peasants' were long vilified throughout the development literature as the scourge of modernization efforts, and assumptions regarding the tragedy of the commons suggested that local responses to stress actually tended to augment risk. But respect is growing for the ways in which local communities diversify production strategies, shepherd scarce resources and cope with natural hazards. Furthermore, in virtually all disasters in the South, it is the local communities that provide the vast majority of support to victims in crucial, acute stages.

Calls for caution have been raised, however, about romanticizing the role of communities in managing local resources, as even these institutions operate in dynamic and often disturbing ecological and political contexts, where short-term survival and long-term risk reduction do not always go hand in hand (Agrawal and Gibson 1999). While community mechanisms may help in dealing with individual tragedies, they are proving increasingly weak in dealing with covariant risks affecting broad sectors of the population (Dercon 2002). Concerns have been raised about general tendencies in the humanitarian discourse to turn over responsibilities to 'the victims themselves', which could be justified by exaggerated faith in the 'community'. Furthermore, traditional local strategies for vulnerability reduction are of little use in the face of new vulnerabilities emerging from changing demographics and economic development trajectories (Ashley and Maxwell 2001). AIDS, the declining profitability of small farms, the inability of the poor to access changing markets, and a general trend for rural populations to depend on non-farm income have had major impact on coping strategies. The poorest have less access to traditional risk reduction and insurance strategies, as age-old institutions become less viable and as the most vulnerable are excluded from effective membership in the 'community'. These communities, furthermore, may have already been torn apart my conflicts, with different factions vying for access to relief and reconstruction resources (Jensen and Stepputat 2001). The implications of stress and demographic changes on these mechanisms has been understood for many years (Scott 1976), but can easily be forgotten when communities are held up as a potential replacement for the state. Even where they work reasonably well, community insurance 'works for some, but not for many' (Goldstein *et al.* 2002: 27). Increasing the security of farming systems may, for example, have little impact on the livelihoods of those who no longer farm, and may even have negative effects if the most vulnerable people are consumers relying on low prices, rather than producers needing protection from cheap imports (Farrington *et al.* 2002).

### *Government*

In many respects the most important of the actors is the government. Despite neo-liberal calls for reducing the role of the state, human rights law still clearly declares that governments bear the ultimate responsibility for the safety of their citizens (Callamard and van Brabant 2002). Aid is far from sufficient to play more than a complementary role in addressing the challenges of risk management. States must include DMP in their own fiscal priorities if they are to live up to their commitments to respect human rights. It has been noted that while rights-based approaches to development provide excellent points of departure for pressuring governments to shoulder the responsibilities that they have formally accepted, rights-based approaches nonetheless provide little guidance for influencing the prioritization processes in which political decisions are made (Conway and Farrington 2002). The human right to safety and survival can be used to put DMP on the political agenda, but it does not necessarily tell us what to do with it once it is there.

If disaster mitigation and preparedness are to become more than recurrent fads, popping up after a catastrophe to quickly fade from the agenda, then it is in local and national political processes that risk must be anchored. Wisner (2001) points out that there is usually a broad gap between the declarations that accompany disasters about reforming institutions and regulatory frameworks, and the political will and capacity to actually carry through these reforms. The cliché that 'disasters are merely a sign of underdevelopment' easily becomes an excuse to subsume the need to mobilize political will to prevent disasters under a broad and diffuse economic development agenda (Rocha and Christoplos 2001).

### *Multi-actor initiatives*

There is surprisingly little interchange among these different actors, but some efforts are underway to establish bridges among different perspectives. ProVention is a global coalition of governments, international organizations, academic institutions, the private sector, and civil society organizations led by the World Bank, the International Federation of Red Cross and Red Crescent Societies and the United Nations Development Programme. It aims to reduce disaster impacts in developing countries. The ProVention Consortium is an example of a mechanism that intends to address the conceptual and operational gaps between actors to manage risk and thereby reposition disasters on the development and humanitarian agendas. Similar efforts are afoot at regional levels. In Central America the IDB, together with the World Bank and various bilateral agencies, is supporting the Centre for Co-ordination of Natural Disaster Prevention (Centro de Coordinación para la Prevención de los Desastres Naturales – CEPREDENAC) as a platform for bringing together state, civil society and donors to address risk.

## Decentralization, service provision and risk

Even if many actors are talking more about risk, there is a danger that this new debate may remain at the level of talk and macro-level infrastructural investment. Efforts to pursue a livelihoods perspective have pointed to the necessity to consider the perspectives, capacities and priorities of poor people themselves and the local institutions with which they interact (Narayan *et al.* 2000). A global agenda of decentralization (of government, the private sector and NGOs) has shifted responsibilities for DMP closer to those affected by disasters. This has been driven by ideologies of both left and right, where distrust of central governments and faith in popular participation have come together in assumptions that risk is best addressed as close as possible to where the disasters actually occur. This agenda has gained further credence through faith that information technology will enable local authorities and NGO offices to pull down the resources they need to cope with risk. Indications are mixed regarding how well local actors can mobilize to shoulder these responsibilities (see Chapter 11 this volume). Positive signs may be found in the growing ownership of regulatory mechanisms for land use and forestry by local government, and in synergy between NGOs and local government (Lele 2000). Nonetheless, decentralization of responsibilities without increased access to financial resources at local level has grave limitations. It may be a hidden way for government simply to shirk its responsibilities.

In the context of decentralization amid weakening public services, a fundamental challenge is to synchronize DMP efforts with the strategies of vulnerable communities. At one level, this means renewed efforts to understand and work with local institutional partners. This involves finding ways to support local service institutions to adapt how they relate to their clients in turbulent contexts. Organizations such as Red Cross/Crescent Societies, religious institutions and farmer organizations have a presence at grassroots level, and are ideally placed to develop an understanding and knowledge of local capacities and coping strategies and, moreover, to involve communities in shaping DMP activities. Experience has shown that their roles are essential, but has also shown that outside agencies have not been effective in supporting their erstwhile 'partners' to play these roles.

As mentioned earlier, local institutional development is not a panacea. The ultimate impact of decentralization on DMP is intimately related to the weakening of state institutions and the changing frameworks of local control of resources and decision-making. These political questions impinge on preparing for disasters since they involve identifying who will be responsible for doing what in an emergency and in related rehabilitation and development initiatives. Structural adjustment and the decline of state control over public services have meant that there is no longer a self-evident meaning to the traditional assumptions that NGOs limit their role to filling temporary gaps in the capacity of the state to provide basic public services. As governments look for opportunities to hand over their established responsibilities to 'civil society', NGOs can easily be pulled into providing services that they cannot sustain. The vacuum in provision of basic public services is most glaring when addressing the implications of providing such services in a crisis. It is impossible to suggest concrete guidelines regarding 'who does what' in the myriad of new configurations which are currently emerging in humanitarian practice and decentralized development. Neither, however, can the division of responsibilities be ignored. Decisions must be made about how to help the most vulnerable, while keeping an open and constructive dialogue going about which agency (the government, NGOs, private sector, etc.) should eventually shoulder the burdens of basic public welfare.

## Triage: sustainability versus risk management?

The marketplace will not mitigate risk and respond to disasters by itself. Civil society will not replace the state. Local civil servants have few incentives suddenly to start addressing disaster hazards alongside their day-to-day work. These different actors can, however, find greater synergy. This depends on political will. Political will depends on exertion of political leadership amid a shifting set of incentives, pressures and polemics. The political costs of redirecting priorities from visible development projects to addressing abstract long-term threats are great. It is hard to gain votes by pointing out that a disaster *did not* happen. On the other hand, disasters as indicators of failed development also provide opportunities for reformers, who can draw attention to the failures of current development models (t'Hart 2001). NGOs play a major role in calling for political commitment, but they have proved to be rather fickle themselves with regards to shifting the focus of their projects, and even their advocacy, according to changing politico-economic circumstances. Addressing the issue of political will requires moving beyond calls for grand technocratic solutions. Instead of focusing on *which* technical choice is most appropriate, we must understand the political process that determines *how* these choices are made.

Why do development actors consistently refuse to engage in a serious manner with risk management? What are the origins of the divide between the discourse on development and initiatives inspired by humanitarian values, such as disaster response and vulnerability reduction? A significant part of the answer lies in the shibboleth of 'sustainability'. There is a refusal to admit that even in the mid- and long-term outside resources will need to be channelled to the most vulnerable. There is no quick fix for the marginalization of the hinterlands that is emerging in the face of globalization. The magic of the marketplace (and even internet) is not showing up at the doorsteps of the poor farmers struggling to plant a few cassava bushes on an eroded hillside in the midst of a civil war. In order to find synergy between humanitarian and development strategies, a more open-minded and flexible approach is needed to channelling public resources into the livelihoods of the poorest and in vulnerability reduction. The belief that a little aid for a few years will lead to self-reliance is totally unfounded in many parts of the world. Safety nets are needed, both for dealing with the chronically poor and for those 'shocked' by disaster (Devereaux 2002; Sumarto *et al.* 2000).

Triage is a useful concept for facing the questions surrounding how to anchor a concern for risk in development efforts. In *Collins English Dictionary* (1991) Triage is defined as 'the principle or practice of allocating limited resources, as of food or foreign aid, on a basis of expediency rather than according to moral principles or the needs of the recipients'. This usage of the term stems from battlefield medicine, where casualties are sorted according to those who will survive without treatment, those who will probably not survive at all, and those in-between for whom treatment will yield greatest impact. Even though 'triage' is a word rarely used in studies of development, it has nonetheless been a de facto guide for many investments. It is a useful way of shedding light on the practical and ethical choices to be made in prioritization of investments, and for placing this prioritization within the broader context of how development policy relates to humanitarian (rather than economic) values. Can a concern with reducing human suffering be combined with promoting economic growth, and if so how?

Trends in rural and urban development in the face of globalization have shown that the need to find a common ground between developmentalism and humanitarian values is more acute than ever. 'Durable disorder' (Duffield

2000) is now taking hold in marginal areas in the form of chronic violence and social alienation. Transnational economic networks are taking advantage of the withdrawal of the state from isolated rural areas (and even no-go zones in the cities) by establishing smuggling, production of narcotics and other forms of illicit enterprise. This phenomenon suggests that there are heavy economic costs (in addition to ethical issues) stemming from conflict, criminality and social disintegration when services are withdrawn. Dismantlement of 'unsustainable development' has proved unsustainable as well.

If triage is to be used as an analytical concept for understanding the choices that actors make, and not as a recipe for exclusion, this will mean bringing risk management into the sphere of the ongoing national and international debates on the respective roles of government, civil society and the private sector. If we look at how policy decisions are being made based on triage, we can then ask 'what happens when vulnerability is ignored?'. The consequences of triage can be made more apparent. Risk management falls too often between the cracks of the policy debate. Calls for rights-based development may highlight failures, but provide little guidance for understanding how priorities can and must be determined. Making the choices and implications of these choices transparent is the first step in identifying where and why the political will may be found for reducing risk.

## References

Agrawal, A. and Gibson (1999) 'Enchantment and disenchantment: the role of community in natural resource conservation', Available online: http://www.mekonginfo.org/mrc_en/doclib.nsf/0/F52F8186F85933F9472568E80044F906/$FILE/FULLTEXT.html (accessed May 2002).

Ashley, C. and Maxwell, S. (2001) 'Rethinking rural development', *Development Policy Review* 19 (4), 395–425.

Bankoff, G. (2001) 'Rendering the world unsafe: "vulnerability" as Western discourse', *Disasters* 25 (1), 19–35.

Benson, C., Twigg, J. and Myers, M. (2001) 'NGO initiatives in risk reduction: an overview', *Disasters* 25 (3), 199–215.

Blench, R. and Marriage, Z. (1998) 'Climatic uncertainty and natural resource policy: what should the role of government be?', *ODI Natural Resource Perspectives* No. 31, April.

Callamard, A. and van Brabant, K. (2002) 'Accountability in humanitarian operations: a question of rights and duties', *World Disasters Report 2002*, Oxford: Oxford University Press.

Carney, D. (ed.) (1998) *Sustainable Rural Livelihoods: What Contribution Can We Make?*, London: Department for International Development.

Charvériat, C. (2000) *Natural Disasters in Latin America and the Caribbean: An Overview of Risk*, Inter-American Development Bank Working Paper No. 434, Washington, D.C.: IDB.

Christoplos, I. (1998) *Sensemaking in Services: Perspectives from the Frontline in Relief and Development Practice*, Agraria 92, Swedish University of Agricultural Sciences, Uppsala.

Christoplos, I. (2000) 'Natural disasters, complex emergencies and public services: rejuxtaposing the narratives after Hurricane Mitch', in P. Collins (ed.) *Applying Public Administration in Development: Guideposts to the Future*, Chichester: John Wiley and Sons.

*Collins English Dictionary* (1991) Glasgow: HarperCollins Publishers.

Conway, T., Moser, C., Norton, A. and Farrington, J. (2002, forthcoming) *Rights and Livelihoods Approaches: Exploring Policy Dimensions*, Natural Resource Perspectives, London: Overseas Development Institute.

Dercon, S. (2002) *Income Risk, Coping Strategies and Safety Nets*, WIDER Discussion Paper No. 2002/22, Helsinki.

Devereaux, S. (2002) *Social Protection and the Poor: Lessons from Recent International Experience*, IDS Working Paper No. 142, Brighton: University of Sussex.

Disasters (2001) Special Issue: 'Politics and Humanitarian Aid: Debates, Dilemma Dissension', *Disasters* 25 (4).

Duffield, M. (1994) 'Complex emergencies and the crisis of developmentalism', *IDS Bulletin* 25 (4), 37–45.

Duffield, M. (2000) 'The emerging development–security complex', in P. Collins (ed.) *Applying Public Administration in Development*, Chichester: Wiley.

Farrington, J., Christoplos, I. and Kidd, A. (2002), *Extension, Poverty and Vulnerability: The Scope for Policy Reform – Final Report*, ODI Working Paper No. 155, London: Overseas Development Institute.

Fowler, A. (2000) *Civil Society, NGDOs and Social Development: Changing the Rules of the Game*, Geneva: UNRISD.

Goldstein, M., de Janvry, A. and Sadoulet, E. (2002) *Is a Friend in Need a Friend Indeed? Inclusion and Exclusion in Mutual Insurance Networks in Southern Ghana*, WIDER Discussion Paper No. 2002/25, Helsinki.

Hewitt, K. (1996) *Regions of Risk*, London: Longman.

Hulme, D. and Edwards, M. (eds) (1997) *NGOs, States and Donors: Too Close for Comfort?*, London: Save the Children.

International Federation of Red Cross and Red Crescent Societies (IFRC) (2002) *World Disasters Report 2002: Forms on Reducing Risk*, Geneva: IFRC.

Jensen, S. and Stepputat, F. (2001) *Demobilizing Armed Civilians*, CDR Policy Paper, Copenhagen: Centre for Development Research.

Kreimer, A. and Arnold, M. (eds) (2000) *Managing Disaster Risk in Emerging Economies*, Washington, D.C.: World Bank.

Kreimer, A., Arnold, M., Barham, C., Freeman, P., Gilbert, R., Krimgold, F., Lester, R., Pollner, J.D. and Vogt, T. (1999) *Managing Disaster Risk in Mexico: Market Incentives for Mitigation*, Washington, D.C.: World Bank.

Lele, S. (2000) *Godsend, Sleight of Hand or Just Muddling Through: Joint Water and Forest Management in India*, ODI Natural Resource Perspectives No. 53, April, London: Overseas Development Institute.

Long, N. and Long, A. (eds) (1992) *Battlefields of Knowledge: The Interlocking of Theory and Practice in Social Research and Development*, London: Routledge.

Narayan, D., Chambers, R., Shah, M.K. and Petesch, P. (2000) *Voices of the Poor: Crying Out for Change*, Oxford: Oxford University Press, for the World Bank.

Rocha, J.L. and Christoplos, I. (2001) 'Disaster mitigation and preparedness on the Nicaraguan post-Mitch agenda', *Disasters* 25 (3), 185–198.

Scott, J.C. (1976) *The Moral Economy of the Peasant*, New Haven, Conn.: Yale University Press.

Skees, J., Varangis, P., Larson, D. and Siegel, P. (2002) *Can Financial Markets be Tapped to Help Poor People Cope with Weather Risks?*, WIDER Discussion Paper No. 2002/23, Helsinki.

Slim, H. (2002) 'Claiming a humanitarian imperative: NGOs and the cultivation of humanitarian duty', Paper presented at the Seventh Annual Conference of Webster University on Humanitarian Values for the Twenty-First Century, Geneva 21–22 February.

SPHERE (2000) 'Code of conduct for the International Red Cross and Red Crescent Movement and NGOs in disaster relief', in *SPHERE Handbook*, Oxford: Oxfam.

Stone, D., Maxwell, S. and Keeting, M. (2001) 'Bridging Research and Policy', Available online: <http://www.gdnet.ids.ac.uk/tm-frame.html? http://www.gdnet.org/pdf/Bridging.pdf.> (accessed May 2002).

Sumarto, S., Suryahadi, A. and Pritchett, L. (2000) *Safety Nets and Safety Ropes: Who Benefited from two Indonesian Crisis Programmes – the 'Poor' or the 'Shocked'?* Policy Research Working Paper No. 2436, Washington, D.C.: World Bank.

t'Hart, P. (2001) 'Political leadership in crisis management: an impossible job?', Keynote address, research meeting 'The Future of European Crisis Management', Uppsala University, 21 March.

Twigg, J. (2001a) *Sustainable Livelihoods and Vulnerability to Disasters*, Benfield Greig Hazard Research Centre, University College London, Available online: <http://www.bghrc.com/> (accessed May 2002).

Twigg, J. (2001b) *Corporate Social Responsibility and Disaster Reduction: A Global Overview*, Benfield Greig Hazard Research Centre, University College London, Available online: http://www.bghrc.com/ (accessed May 2002).

United Nations Development Programme (UNDP) (1999) *Disaster Reduction and Recovery Programme Progress Report and Plan of Action*, Geneva: UNDP Emergency Response Division.

van Aalst, M.K. and Burton, I. (2000) 'Climate change from a development perspective', in A. Kreimer and M. Arnold (eds) *Managing Disaster Risk in Emerging Economies*, Washington, D.C.: World Bank.

Varangis, P. (2001) *Hedging Your Bets*, Available online: <www.worldbank.org/html/fpd/dmf/weather.htm> (accessed May 2002).

Wisner, B. (2001) 'Risk and the neo-liberal state: why post-mitch lessons didn't reduce El Salvador's earthquake losses', *Disasters* 25 (3), 251–268.

World Bank (2000) *World Development Report 2001*, Oxford: Oxford University Press.

# 7 Beyond disaster, beyond diplomacy

*Ilan Kelman*

## Introduction

Disaster diplomacy began with Kelman and Koukis (2000: 214) asking: 'do natural disasters induce international cooperation amongst countries that have traditionally been "enemies"?'. They implied that local or regional disasters could positively affect bilateral relations amongst states which would not normally be prone to such co-operation, and that although disasters are felt at the local scale they might stimulate political co-operation at an international level. The potential for a locally situated disaster to affect international politics far beyond the physical reach of the disaster event is referred to here as a potentially globalizing effect of disaster. A similar but temporal disjunctive exists in the often fleeting interest of actors in disaster and diplomacy, compared to the longer-term gestation of root causes.

This chapter explores the disaster diplomacy thesis through several case studies. Relationships are examined between the political levels at which disaster diplomacy operates, and the local, national and global pressures that shape whether or not the potential for disaster diplomacy is realized. The challenge is to counteract an inherent contradiction between disaster and diplomacy cause and effect, where disaster diplomacy tends to become 'globalized' over a short time-frame while the local and global root causes of disaster often remain over the long-term. The unanswered question is whether we could predict when a specific disaster event might be a catalyst for international disaster diplomacy.

## The immediate disaster diplomacy family

The principle behind exploring disaster diplomacy was that:

> The occurrence or threat of disaster creates opportunities to facilitate better cooperation or relations amongst states in conflict through fostering linkages which otherwise might not have existed. The cooperative spirit generated from common efforts to deal with disasters – through either perceived necessity or choice from the humanitarian imperative –

possibly overrides pre-existing prejudices, breaking down barriers which then may never be rebuilt.

(modified, with author permission, from Kelman and Koukis 2000: 214)

Discussion has widened, with a growing set of case studies and theoretical analyses (http://www.disasterdiplomacy.org). Three case studies, covering different disaster types in different regions, were the original exploration into disaster diplomacy. This trio is presented and updated to open the discussion and is followed by two further case studies.

### *The 1999 earthquakes in Greece and Turkey*

On 17 August 1999 an earthquake killed more than 17,000 people across north-western Turkey. The Greek government and people responded with sympathy, money, goods and rescue workers only to have Athens struck by an earthquake on 7 September 1999 which killed more than 140 people. Turkey, still struggling in the aftermath of their disaster, responded as Greece had done with initiatives from leading politicians as well as the populace. At least five further fatal earthquakes hit Turkey from 31 August 1999 until a sixth earthquake killed at least 894 people on 12 November 1999.

The rapid thaw in Greek–Turkish relations following the earthquakes prompted Ker-Lindsay to explore the influence of these disasters on this diplomacy (in Kelman and Koukis 2000). He noted that the rapprochement process had started earlier in 1999 in the wake of NATO military action in Kosovo and Serbia. The earthquakes thrust the process into the spotlight and achieved far more than was predicted from the pre-earthquake talks. The disaster, though, was clearly not the root cause of the diplomacy. Ker-Lindsay even suggested that the process may have been hindered by forcing difficult diplomacy into the public's eye, thereby raising expectations and providing targets for critics.

Subsequent events in Greek–Turkish relations have borne out the concerns of Ker-Lindsay. For example, on 22 March 2001 the Greek Defence Minister Akis Tsochatzopoulos, generally perceived as hawkish, stated not only that Greece and Turkey should be co-operating more but also that they had a responsibility to do so for the stability of the region. Ankara's response was lukewarm. While many commentators were disappointed, in November 2001 Greece and Turkey jointly sponsored a UN resolution on dealing with natural disasters and, during February 2002, they started high-level talks on Aegean issues.

Relations in Cyprus between the two governments on the island have also improved, with the Cypriot president and north Cypriot leader holding a series of meetings in January 2002. Such complex Greek–Turkish–Cypriot diplomatic interactions are as much a result of Balkan conflicts – including the insurgency by ethnic Albanians in Macedonia throughout 2001 – and

Cyprus' wish to join the European Union peacefully in 2004 as from any spillover due to the 1999 earthquakes.

### *Cuba, the USA and climate-related disasters*

Glantz discussed the impact of climate-related threats on American–Cuban relations (in Kelman and Koukis 2000). He discussed the entire range of disaster management activities, including scientific co-operation in researching long-term climatic trends and real-time monitoring of tropical cyclones. Most co-operation appeared to occur through individual contacts at the scientific level, whereas high-level, publicly visible connections were discouraged. The conclusion was that climate-related disaster management forges linkages between Cuba and the USA, yet disaster issues alone will not resolve the political difficulties. Glantz also noted with respect to emergency relief that 'at some levels in the US government, there is appreciation that disaster losses can destabilise regimes, so why help the [Fidel] Castro regime stay in place?' (Kelman and Koukis 2000: 242).

With respect to the post-disaster actions of relief and recovery, Glantz used the devastating drought of 1998 in Cuba to explain the animosity which exists between the two countries. Cuba requested food assistance from the UN but 'would not accept any aid that obviously came from the US, arguing that food aid was needed in the first place as a result of the US embargo and not because of the drought alone' (Kelman and Koukis 2000: 242).

This pattern looked repeatable in early November 2001 when Hurricane Michelle became the most powerful hurricane to cross Cuba during Castro's reign. Excellent pre-disaster preparation and a population obedient to evacuation orders minimized casualties. Five deaths occurred in Cuba (population 11 million) in contrast to the approximate toll of twenty from Nicaragua and Honduras (combined population 12 million). Nonetheless, extensive damage occurred in Cuba and international assistance was needed. Washington offered assistance on 7 November 2001, which Havana declined while offering to pay for the goods as long as Cuban ships transported them. Washington refused this counteroffer. Diplomatic dancing eventually led to American food purchased by Cuba arriving in Havana on an American ship on 16 December 2001.

Did this disaster diplomacy originate from only Hurricane Michelle's destruction? Or was behind-the-scenes manoeuvring present before the hurricane, which the disaster forced into the public spotlight as with the earthquakes in Greece and Turkey? One possibility is that the horrific terrorist attacks of 11 September 2001, during which American commercial jets were hijacked and flown into the World Trade Center and the Pentagon, jolted traditional enmity in international politics. Washington may have learned that Americans' 'usual enemies' feel compassion for and can support the USA when faced with heinous atrocities, while concern grew from new perceived threats from international terrorism. This phenomenon may be a form of disaster diplomacy displaying itself through the disaster of 11 September. The lessons which Cuba may have learned or be expecting to give in initially turning down the ostensibly conciliatory gesture from the Americans are not clear.

### *Drought in southern Africa, 1991–1993*

Holloway (in Kelman and Koukis 2000) discussed the 1991 to 1993 drought in southern Africa. Against a background of massive political change across the region, southern Africa initiated and co-ordinated the largest international import of food to counter a drought since 1966–1967 in India. Holloway examined the interplay between these simultaneous political and humanitarian success stories to conclude that 'while diplomatic dividends can indeed flow from disaster relief efforts, in this instance, joint cooperation was only possible once potential military, economic, and other forms of regional confrontation that dominated the 1980s had been controlled' (Kelman and Koukis 2000: 273).

Holloway aptly illustrates that starvation during drought is often due to politics as much as rainfall variability. Many drought emergencies are marked by food being widely available locally, yet simply unaffordable (Shawcross 2000, quoting a conversation with Fred Cuny on Cuny's experiences, starting with the Biafran crisis in 1968). In Holloway's case, agricultural vulnerability afforded opportunities to decrease political vulnerability, an interaction which prevented rather than produced a disaster.

Why was this case different? The unique political situation seems to have had the most influence: the political climate of transition and peace in the southern African community at that time was conducive towards collaboration. This pattern perhaps emerges more frequently, but our bias towards Africa as a crisis zone obscures our vision. Irrespective, it is apparent that, as with the other two case studies, the diplomacy had to be present for disaster management to reap further diplomatic benefits.

### *India/Pakistan*

On 26 January 2001 an earthquake hit western India, killing more than 20,000 people. Pakistan suffered some damage too, but Gujarat, India was devastated. Almost immediately, Pakistan offered assistance. On 2 February 2001 India's Prime Minister Vajpayee and Pakistan's military leader and de facto ruler General Musharraf (for background on Musharraf's rise to power and political role at the time, see http://news.bbc.co.uk/hi/english/world/south_asia/newsid_1742000/1742997.stm) spoke on the telephone, the first time they had had direct contact – at least, in public. The diplomatic outcome included a summit of the two leaders in India from 14 to 16 July 2001.

On 20 June 2001, just prior to the summit, General Musharraf deposed Pakistan's figurehead president. The purposes seemed to be to consolidate his power and to give himself a more prestigious position – above that of Prime Minister Vajpayee – for the upcoming negotiations. Such action was hardly likely to encourage trust or to indicate a commitment to peace and democracy. These ideas were probably not foremost in President Musharraf's mind, implying that his actions and sudden interest in diplomacy may have had underlying reasons other than to help a neighbour suffering disaster.

Despite, or because of, high hopes and intense scrutiny, a final statement on the summit to be signed by both leaders could not be agreed. Some successful initiatives emerged, but commentators generally agreed that the meeting failed. Nonetheless, both leaders seemed genuinely to have sought common ground. This process could have built trust and permitted further moves towards long-term peace. Vajpayee agreed to visit Pakistan by the end of 2001, but, amid rising tensions over Kashmir, the visit did not occur. In the following months, Indo-Pakistani relations alternated between near-war and thaws, but optimism from the 'earthquake diplomacy' of a year earlier had dissipated.

The earthquake triggered closer ties between Pakistan and India, which could have yielded, and may still yield, longer-term, positive diplomatic outcomes. A devastating earthquake has not been enough for resolving this conflict.

### *North Korea*

Since the end of the Korean War in 1953, North Korea has been internationally isolated. Starting in about 1995, a famine induced by a series of floods and droughts, and likely compounded by agricultural and economic mismanagement, severely disrupted the country and reportedly killed more than one million people. International relief operations navigated the political minefield to assist while, simultaneously, North Korea began to re-enter the international community. On 7 March 2000, Japan sent food aid to North Korea as part of a deal for starting diplomatic talks. The new era in Korean relations culminated on 15 June 2000 when North Korean President Kim Jong-il and South Korean President Kim Dae-jung met in Pyongyang.

Did the floods, droughts and famine contribute to North Korea's emergence? Kim Dae-jung was elected President of South Korea in December 1997 promising to bring a new era to relations with North Korea – which he obviously did. Even if his promises were not influenced by news of the humanitarian crisis trickling out from North Korea, his successes may have stemmed from the need for aid. The large scale of the disaster might be significant, considering that other disasters had little influence on Korean relations. Severe storms and storm surges affect the west coast of the Korean peninsula (Kim *et al.* 1998) but have not produced diplomatic results.

Irrespective of any diplomacy due to the famine, the North's response to Kim Dae-jung's election was frequently to push his tolerance through incursions into the South and missile testing. Kim Dae-jung winning the 2000 Nobel Peace Prize provoked a harsh response from Pyongyang, indicating that disaster would not be the overriding influence in North Korea's interaction with the outside world. Nonetheless, food aid was used to force diplomatic issues as part of a complex series of negotiations involving aid, diplomatic ties and military capability. Without the need for aid, the other issues may not have arisen.

The potential longevity of any disaster diplomacy for North Korea was shortened by American President George W. Bush. In the first months of his Presidency in 2001 he preferred an ideological foreign policy unilaterally unleashed upon the world without regard to any humanitarian imperative. Following the terrorism of 11 September 2001, an event local to the USA, the Bush administration's foreign policy considered and involved the international community to a greater extent, although marked insularity was still evident. Some dividends may have emerged regarding Cuba, as noted previously, but President Bush termed North Korea 'evil' in early 2002. The famine produced regional diplomatic opportunities which were grasped, but then dwarfed at the international scale by other considerations, including domestic American politics.

### *A larger family of disaster diplomacies*

Other possible disaster diplomacy case studies abound. Israel's Ministry of Foreign Affairs summarized Israeli humanitarian relief operations between 1997 and 1999, which included work in Afghanistan, Azerbaijan and Russia (MFA 2002). The Middle East Regional Cooperation Programme aims to provide data for the implementation of seismic building codes in Israel, Jordan and the Palestinian National Authority (http://www.relemr-merc.org). Diplomatic outcomes from either project have not been unambiguously identified, but they use, respectively, response and mitigation across borders which have not normally had straightforward, overt political co-operation.

On disaster rather than geographic themes, the World Health Organization runs disease control programmes which require co-operation from all countries (http://www.who.int/aboutwho/en/disease_er.htm). The Global Seismic Hazard Assessment Program (http://seismo.ethz.ch/gshap), launched in 1992, examined test areas in the conflict zones of the Caucasus and India–China–Tibet.

The search for Near-Earth Objects (NEOs, including asteroids and comets which may strike the Earth; see http://impact.arc.nasa.gov and http://www.nearearthobjects.co.uk) could form the basis for linkages amongst countries that normally would not consider working together. When political decisions on NEOs must be made – for example, related to monitoring NEOs, planning for a crisis, preventing a strike, or responding to a strike

– these scientific linkages have the potential for leading to interaction at higher political levels. A global disaster, such as a NEO flattening a major city's centre or causing a massive tsunami, could forge long-term linkages too. One unknown in NEO disaster diplomacy would be the response from Washington, with its strong weapons capability for attempting to deflect an incoming NEO, if the threatened city were Kim Jong-il's Pyongyang or Castro's Havana rather than Boston or San Francisco – or if the NEO's trajectory were so uncertain that Miami and Castro's Havana were equally threatened.

Two final examples illustrate the challenges yet to be overcome by disaster diplomacy. From 1998, Ethiopia and Eritrea fought a border war that killed thousands of soldiers. In early 2000, a drought afflicted Ethiopia, which had previously relied on Eritrean ports for importing goods. Aid workers attempted to open a humanitarian corridor through Eritrea to Ethiopia, achieving only minor success.

China and Taiwan offer another 'non-success', if not a complete failure. An earthquake killed more than 2,000 people in Taiwan on 21 September 1999. China offered sympathy, money and goods. This gesture was appreciated by the Taiwanese, but they nonetheless remained suspicious of China's motives in trying to assert control over Taiwan's affairs in times of need. This mistrust was exacerbated by the Chinese Red Cross attempting to become the first point of contact for aid rather than the Taiwanese Red Cross.

Some diplomatic successes resulted, such as on 6 February 2001 when the first Chinese boat in 50 years officially visited Taiwan. Several events since the earthquake, however, affected relations between China and the USA, and thus the issue of Taiwan, as much as the earthquake. For example, on 1 April 2001, an American spy aeroplane collided with a Chinese air force aeroplane killing the Chinese pilot and forcing the American aeroplane to make an emergency landing in China. As well, the visit of Taiwan's president to New York (21–23 May 2001) coincided with President Bush meeting the Dalai Lama in the White House (22 May 2001), evoking strong condemnation from Beijing. Despite the possibility of a post-earthquake thaw, circumstances were not conducive to the disaster having a lasting legacy on this diplomacy.

## Implementing the disaster diplomacy agenda

### *Passive and active disaster diplomacy*

The case studies indicate that disaster diplomacy has a tangible, but not an overriding, influence on diplomatic efforts. Disaster management can significantly influence diplomatic processes which have already started but cannot by itself induce new diplomatic initiatives. Any gains are easily lost through the influence of other events, including further disasters. Disaster can act as a catalyst, but not as a creator, of diplomacy.

Why this catalyst manifests in some cases and not in others is not yet understood. More importantly, if we see that disaster could have a positive impact on diplomacy, would methods exist for ensuring that the catalytic effect is made to appear? The case studies have described events when disaster management has interacted with international affairs without the players necessarily being aware of how to manipulate the interactions for positive gain. In effect, they are examples of passive disaster diplomacy.

Now that the role which disaster could play in international affairs is recognized, deliberate application must be considered. Through active disaster diplomacy we could maximize opportunities (for example, the 2001 Gujarat earthquake or Hurricane Michelle) to ensure that disaster events contribute to the long-term easing of tensions amongst antagonistic states.

Active disaster diplomacy may involve working with the media or lobbying governments to raise awareness of the potential for improving international relations on the basis of disaster-related co-operation. We could contend that disaster management can and should form a basis for breaking down diplomatic barriers. Perhaps disaster management forms the best such basis since it concerns the fundamental right to life. In a post-disaster scenario especially, compassion and the humanitarian imperative derived from the populace could be used to encourage hostile governments to move towards reconciliation.

But we also need to exercise caution. It is unrealistic to expect that disaster diplomacy can influence all international political relationships. Pushing too hard for disaster diplomacy in some cases might distract efforts away from more promising peace and development processes. To determine the most effective mechanism for countering disasters in politically volatile regions, an awareness of what disaster diplomacy can and cannot do is needed.

## The extended disaster diplomacy family

### *Expanding disaster diplomacy*

The original definition of disaster diplomacy was strict. Natural disaster, as an event or situation which overwhelms a community's coping ability, covered the entire disaster management cycle: pre-disaster actions such as prevention and preparation and post-disaster actions such as response and recovery. The scope, though, covered only immediate, obvious threats, such as tropical cyclones, droughts, earthquakes, epidemics, floods, NEOs and volcanoes.

Pelling (2001) labels such events as 'catastrophic' disasters in contrast to 'chronic' disasters, which still overwhelm a community's ability to cope yet which manifest every day. Examples are a low-quality water supply, a non-sustainable and inadequate energy supply and poor waste management. These chronic disasters lead to a lower health standard and poorer environmental quality. The lack of resources to invest in sustainable cultivation practices or to tackle outside interests exploiting local human and material resources are further examples alongside 'the gendered and ethnic nature of social systems' (Pelling 1999: 259). Catastrophic and chronic disasters are not two extreme, exclusive groupings, but fuzzy clusters with similarities.

In parallel, diplomacy must be considered beyond the management of international or bilateral relations by negotiation to cover the entire range of political pressures and responses. Finally, disaster diplomacy asked only how disaster influences diplomacy, yet as noted later, diplomacy affects disaster too. Opening disaster diplomacy to the full range of potential interactions between all disasters and all international, political processes permits the development of a thorough understanding of the role of disaster. One implication is that rather than all outcomes being positive, as was the original intent of examining disaster diplomacy, negative interactions may result.

### *Natural disasters as weapons of war*

At least three cases exist of Russia's winter weather being a significant influence on the defeat of an invading army, although other factors were necessary in each case: Charles XII of Sweden from 1708–1709, Napoleon Bonaparte at the end of 1812 and Adolf Hitler from 1941–1943. Similarly, the victory of the English navy in the Camperdown campaign in 1797 has been attributed as much to weather as to military tactics (Wheeler 1991). These instances describe the inadvertent use of natural hazards to achieve military aims.

Deliberate use of or creation of 'natural' disasters has also been documented. Wagret (1968) describes how parts of the Netherlands were deliberately flooded in 1672 to create a barrier to the invading soldiers of Louis XIV, a technique repeated by both sides during the Second World War to hinder each other. Glantz (in Kelman and Koukis 2000: 246) describes American attempts at cloud seeding over South-east Asia to create rainfall which would disrupt Vietcong activities. Cuba feared that this approach could be used in the Caribbean and, as noted previously, the Americans were not averse to hoping that drought would destabilize Castro. No recorded efforts of inducing earthquakes or volcanic hazards against an adversary have been discovered, but artificial inducement of earthquakes has involved small nuclear explosions and creating reservoirs over fault lines (Waltham 1978).

Lewis (1999: 25) describes two examples of natural disaster catalysing national movements towards political change: 'The November 1970 cyclone, and its subsequent alleged mismanagement, was one of the many influences that triggered the Bangladesh War of Independence which commenced in March 1971 . . . The earthquake that destroyed much of Managua in 1972 triggered in its aftermath the armed uprising [and] ensuing civil war' in Nicaragua which overthrew the government in 1979. Glantz (2001a) suggests that the forest fires which blanketed South-east Asia in haze in 1997 added to the pressures on the Indonesian government. This event spawned a Regional Haze Action Plan (Brauer and Hisham-Hashim 1998); ironically, the same disaster resulted in potential disaster diplomacy amongst South-east Asian countries alongside increased instability at the national level in Jakarta.

Natural disaster and political instability are further intertwined by the notion of 'climate-related flashpoints' as 'a catalyst to change' Glantz (2001a: 1). Climate and climate-related factors must be considered a potential source of political instability within or amongst countries, along with, for example, economic, social and historical factors. A rural population displaced by drought and famine could enter their country's capital and demand compensation, thereby sparking wider unrest or, as environmental refugees, could migrate to and destabilize a neighbouring state.

Political corruption or incompetence may also prevent prevention, resulting in natural hazards becoming disasters. Floods in the UK (Kelman 2001) and India (Pelling 2001) are examples of entrenched political systems and attitudes producing this effect. Famine in Somalia in the 1990s illustrates political instability being a dominant factor in the transformation of climatic variability into a disaster (Shawcross 2000). Glantz (2001b), however, cautions that climatic hazards are often identified as causing political events, such as El Niño abetting the Spanish defeat of the Incas in the sixteenth century and inciting the 1789 French Revolution. Many such attributions may be legitimate speculation from useful research, rather than definite, accurate history.

These examples consider only the 'during' and 'after' phases of a disaster. If war, self-determination or social justice rise from the rubble of a natural disaster, as in Bangladesh and Nicaragua, could similar consequences emerge from the threat of disaster or the existence of a natural hazard? Glantz (2001a) queries whether early warning systems might be feasible and desirable for climate-related flashpoints. If an early warning system indicates that an oppressive regime may fall due to an environment-related flashpoint, should a warning be issued to the threatened government? Since the developed countries tend to have the resources to set up and operate such systems while developing countries tend to have the worrying instabilities, this political engineering would be a Machiavellian form of disaster diplomacy.

### *Local vs. global*

Disaster diplomacy tends to privilege cross-border events above disasters located within a single nation-state. An earthquake or flood straddling an international political boundary does not imply that a larger area or more people are affected than by a similar event in the centre of a country. Instead, society's potential unwillingness to treat both sides of a border as similar before or after a disaster often augments vulnerabilities.

Vulnerability is frequently considered to be predominantly a local condition, with efforts tackling vulnerability taking place at a local level (Lewis 1988). Disaster diplomacy operates at an international or potentially global level. Do dangers exist in forcing a top-down view onto an inherently local concern? In the Gujarat earthquake, the lack of local enforcement of building codes killed thousands. When Washington, London and Moscow became involved in facilitating relations between Islamabad and New Delhi against the backdrop of concerns in Kabul, Dushanbe, Tehran and Beijing, were local vulnerabilities in Gujarat adequately addressed?

Perhaps a genuine hope exists that globalizing local vulnerability – that is, making local vulnerability impact global social and political issues – will ultimately yield the desired rewards. Peace and stability could provide the resources and political will for appropriate, local development activities to succeed.

The melding of these space scales may not be so contradictory. Vulnerability may be local, but natural hazard risks can extend over a wide area; for example, volcanic ash or a tropical cyclone. Thus, disaster risk need not be conceptualized as a local phenomenon. In Autumn 2000, for example, the UK experienced a national flood disaster rooted in an extreme rainfall event, poor planning, mismanaged infrastructure and social unawareness. Each flooded locality had vulnerability characteristics specific to its situation, including blocked drainage ditches, poor cultivation practices, building houses in flood-prone areas and ignoring social vulnerability when building structural defences (Kelman 2001). The outcome in each location was the same: flooded properties, disruption of livelihoods and stressed people. Different local vulnerabilities subjected to the same regional natural hazard produced a similar disaster at each locality, which summed to a national flood disaster (Kelman 2001).

Local vulnerability therefore can lead in aggregate to larger-scale problems and is affected by larger-scale influences. Investigating disaster diplomacy assists in examining one form of interaction amongst space scales. Thus, it is more a dichotomy than a contradiction. The advantage is that international issues have a higher profile than local issues amongst national and international policy-makers. The prominence given to some disaster diplomacy cases provides a two-way result: disaster from local vulnerability conditions positively impacts international problems while the international scale may assist in alleviating the local vulnerability condition.

The Gujarat earthquake continued to receive international publicity a year after the event. A drought in the same region at the same time received far less. The tropical cyclones which hit Gujarat on 9 June 1998 (more than 10,000 fatalities) and Orissa in eastern India on 17 October and 29 October 1999 (more than 10,000 fatalities) are almost forgotten. Vulnerability as an inherently local condition can have either little or significant influence at the international level.

A further concern is that vulnerability has an insidious pattern of crawling between spatial scales. Focusing on developed countries, Blong (1997: 24) writes: 'I suspect that our individual (and hence national) self-reliance has declined alarmingly over the last four or five decades . . . many of us would not be self-reliant in terms of food for the 72-hour period prescribed in all the best disaster manuals.' Choices by relatively rich individuals have scaled up to become a national problem. In the opposite direction, the previous section demonstrated that vulnerability to international conflict may result in natural hazards being used to affect local populations. Vulnerability as an inherently local condition can be scaled up to or trickle down from the international level.

### *Implementing beyond disaster diplomacy*

Much of the prior discussion is based on events as reported by or analysed in the media. Questions are asked and concepts are presented without considering the practicality of enacting thorough investigations into the issues. The detailed, first-hand evidence required to support some of the contentions made in this chapter would need to come from decision-makers while decisions are being made. In the midst of a crisis, is it practical and ethical to attempt to collect such evidence? In order to ascertain motives, if disaster diplomacy processes were articulated to a leader trying to decide on an action, might the decision be influenced; that is, would the observer influence the observed? Would it make a difference to a researcher's methodology if enquiring would influence a leader to increase their rival's vulnerability rather than to provide assistance?

If essential disaster diplomacy evidence is uncollectable, practically or ethically, is insightful exploration worthwhile? The potential exists for numerous academic studies on the topics broached here. Therefore, the potential exists for academic study to interfere, unpredictably and possibly detrimentally, in active processes of political decision-making, disaster response, rapprochement and development. Should 'political testing' – real-time research on tinkering with human societies – undergo as stringent requirements as animal testing (for example, Home Office (2002) details the UK procedures) or experimentation on human subjects? Implementing beyond disaster diplomacy is fraught with as many ethical difficulties as disaster diplomacy processes themselves.

## Conclusions: implementing well beyond disaster diplomacy

Disaster diplomacy has established that disasters can have a catalytic effect for major changes in international affairs, but that disasters do not generate new outcomes on their own and may, in the end, achieve little. Expanding the definition beyond the strict 'disaster' question towards 'development' questions indicates that vulnerability processes rarely generate major political changes themselves but may catalyse them. A political situation may, in fact, stifle changes from disaster management which would have had far-reaching effects otherwise.

How far do such conclusions extend? Disaster has segued into development, which operates in parallel with environmental management. The Antarctic Treaty System (ATS), for example, brings together all interested parties in a forum uninfluenced by conflicts external to the system. Argentina and the UK sat in ATS meetings and negotiated during the Falklands War in 1982. North Korea, despite its reclusiveness, joined as an Acceding State (non-voting member) on 21 January 1987, although has participated minimally so far.

Have positive diplomatic results occurred outside the ATS area, which covers south of 60°S? If so, the lessons of this 'environmental diplomacy' need to be understood in order to apply them to other and future environmental and development treaties. If not, as with disaster, could international environmental management regimes contribute more to the international community than they currently do?

Outer space and the deep sea are two other regions with strong possibilities for international co-operation leading to diplomatic results, or to a war of mass destruction on the earth's surface. The obvious difference is that environmental management regimes are generally welcomed, irrespective of spin-offs, whereas disaster events, unless used as weapons, are normally undesired even though benefits are sought from the destruction. Global climate change is an intriguing issue due to its mix of disaster, development, environmental management and international politics in a morass of social and environmental change at space scales from local to global (see, for example, Newell 2000 and Lewis 1999).

Yet international management of Antarctica, outer space and the deep sea is globally driven. No locals exist (as far as we know) and physical distance from each location does not correlate well with a state's presence there. In contrast, vulnerability of locals can lead to local disaster which, in disaster diplomacy cases, produces global social and political pressures.

The discussion on disaster diplomacy began with the effects of local or regional disaster management on inter-state relations. Expansion beyond the original concept encompassed disaster beyond an immediate, obvious threat and international politics beyond state interactions. One globalized outcome is considering the international community's response to global disaster management, such as climate change which so far has produced more bickering than disaster diplomacy results. Other possibilities for global disasters are the NEO scenarios discussed previously or Ward and Day's (2001) description of a tsunami inundating thousands of kilometres of Atlantic coastline, subjecting cities to a 50-metre high wave.

Could global disaster or local vulnerability be used to drive positive global change in international affairs? Will global environmental change parallel global social change, positively or negatively? Are we, as authors and readers, simply observers or can we effect change? Do observing and commenting themselves effect change? Disaster diplomacy could be a tool or a weapon encapsulating shifts in global politics and international development due to local environmental change and local vulnerability. This approach is highly speculative, but if we refuse to have imagination we are hardly qualified to confront the horrors of disaster and underdevelopment which are often worse than we wish to imagine.

## References

Blong, R. (1997) 'A geography of natural perils', *Australian Geographer* 28, 1, 7–27.

Brauer, M. and Hisham-Hashim, J. (1998) 'Fires in Indonesia: crisis and reaction', *Environmental Science and Technology (News and Research Notes)* 32, 17 (1 September), 404A–407A.

Glantz, M. (2001a) 'Climate-related flashpoints: a useful notion for early warning?', *ENSO Signal* No. 18 (August), 1.

Glantz, M. (2001b) 'Attribution', *ENSO Signal* No. 19 (November), 1.

Home Office (2002) Information provided by the Animal Procedures Section, UK Home Office, Available online: <http://www.homeoffice.gov.uk/ccpd/aps.htm> (accessed 30 January 2002).

Kelman, I. (2001) 'The autumn 2000 floods in England and flood management', *Weather* 56, 10 (October), 346–348, 353–360.

Kelman, I. and Koukis, T. (eds) (2000) 'Disaster diplomacy', Special section in *Cambridge Review of International Affairs* XIV, 1, 214–294.

Kim, S.-C., Chen, J., Park, K. and Choi, J.K. (1998) 'Coastal surges from extratropical storms on the west coast of the Korean peninsula', *Journal of Coastal Research* 14, 2, 660–666.

Lewis, J. (1988) 'On the line', *Natural Hazards Observer* 12, 4, 4.

Lewis, J. (1999) *Development in Disaster-prone Places: Studies of Vulnerability*, London: Intermediate Technology Publications.

Ministry of Foreign Affairs (MFA) (2002) 'Summary of Israeli humanitarian relief operations: 1997–1999', Israel Ministry of Foreign Affairs (MFA), Available online: <http://www.mfa.gov.il/mfa/go.asp?MFAH0e8v0> (accessed 29 January 2002).

Newell, P. (2000) *Climate for Change: Non-state Actors and the Global Politics of the Greenhouse*, Cambridge: Cambridge University Press.

Pelling, M. (1999) 'The political ecology of flood hazard in urban Guyana', *Geoforum* 30, 249–261.

Pelling, M. (2001) 'Natural disasters?', in N. Castree and B. Braun (eds) *Social Nature: Theory, Practice and Politics*, Oxford: Blackwell.

Shawcross, W. (2000) *Deliver Us from Evil: Warlords and Peacekeepers in a World of Endless Conflict*, London: Bloomsbury.

Wagret, P. (1968) *Polderlands*, London: Methuen and Co.

Waltham, T. (1978) *Catastrophe: The Violent Earth*, New York: Crown Publishers.

Ward, S.N. and Day, S. (2001) 'Cumbre Vieja volcano – potential collapse and tsunami at La Palma, Canary Islands', *Geophysical Research Letters* 28, 17, 3397–3400.

Wheeler, D. (1991) 'The influence of the weather during the Camperdown campaign of 1797', *The Mariner's Mirror* 77, 1, 47–54.

# 8 The insurance industry

## Can it cope with catastrophe?

*Julian E. Salt*

### Introduction

Climate change is happening, but the rate of change is unknown. The key to the climate change debate hinges on emissions of greenhouse gases and their reduction. Globally, humanity emits 6 billion tonnes of carbon into the atmosphere. Since the 1960s the number of large catastrophic events related to weather have increased almost exponentially. The insurance sector (along with the banking sector) will be amongst the first of the sectors to feel the full impacts of climate change.

The objective of this chapter is to assess climate change issues and to lay out potential solutions to the problem, especially where engagement by the insurance industry is relevant. In general, most of the actors in the insurance chain have little or no real perception of climate change issues and do not take them into account when pricing insurance policies for property. The increased levels of risk from natural events has to be factored into new insurance products, otherwise there exists a potential for a future massive and unsustainable loss from climate-induced events.

The debate over climate change has moved through several phases over the years. The respective questions driving the debate have been the following:

1. Is the climate changing?
2. What is causing the change (humanity or nature)?
3. How quickly is the climate changing?
4. What impacts can we expect and when?
5. How much will the impacts cost?
6. Who will pay?
7. What can be done to slow down or reverse the trend?

Some of these questions impact directly on the insurance and financial industry, while others have a more indirect effect, albeit nonetheless important. The objective of this chapter is to explore these seven questions and lay out a potential plan to solve the problem.

## Is the climate changing?

Ever since the Intergovernmental Panel on Climate Change (IPCC) was created in 1988 a steady stream of potential impacts from climate change has been issued to the world (see Chapter 2 this volume). The proxies for climate change include such factors as concentration of $CO_2$ in the atmosphere, global-average air temperature, sea-surface temperatures and sea-level rise. The latest predictions from the IPCC project a potential temperature rise of between 1.4–5.8 °C by 2100 (highly dependent on fossil fuel use), with an associated increase in sea-levels of up to 0.5 m by 2100.

In the UK the Climate Impacts Programme (UKCIP) has been busy creating potential future climate scenarios for the UK over three timescales (2020, 2050 and 2080). Meteorological factors such as air temperature, wind speed, precipitation, humidity, cloudiness, amongst others, have been projected for the next hundred years.

The USA's National Oceanic and Atmospheric Administration (NOAA) have recently announced that globally 1999 was the fifth hottest year on record, 1998 being the hottest ever. Most of the largest warming seems to have occurred over the continental USA land mass. One of the projected impacts of future climate change in the USA and elsewhere is for an increasingly active hydrological cycle, that in turn is predicted to lead to more intense thunderstorms and flash floods. In contrast, increased air temperatures over land (up to 2–3 °C warmer than over the oceans) will bring about drought-like situations that could result in increased forest and bush fires in the more arid parts of the world and increased subsidence conditions in the more urban regions. Water stress for parts of the world such as the Middle East will become evermore apparent.

## Global issues

The key to the climate change debate hinges on the emission (and reduction) of greenhouse gas (ghg) emissions. The present concentration of carbon dioxide (the principle greenhouse gas) in the atmosphere stands at 365 ppm and is rising at 1 per cent/yr. If things remain unchecked the concentration of $CO_2$ in the atmosphere will race to 550 ppm (twice pre-industrial levels) and higher, resulting in a potential runaway greenhouse effect.

In the worst-case scenario, positive feedbacks in the planet's weather system could mean that ice-caps partially/fully melt, resulting in the partial/total shut-down of the Atlantic Conveyor (Gulf Stream effect) which could bring a mini ice age to Europe. Thawing of Russian permafrost and undersea methane hydrates could result in a positive runaway in emissions of methane (second most important ghg). Also, extensive die-back of the equatorial forests, to the extent that Amazonia could become a net source of carbon emissions as opposed to its present net sink characteristic, is a real possibility once certain threshold concentrations of $CO_2$ in the atmosphere have been reached.

Globally, humanity emits 6 billion tonnes of carbon into the atmosphere every year. By a strange quirk there are 6 billion people on the planet. Thus, in an ideal world, this would mean that everyone should be allowed to emit 1 tonne of carbon per year. However, the world is not a fair place, as evidenced by the fact that the developed world emits far beyond their per capita share of greenhouse gases, while the poorer nations emit way below their theoretical allowance. In an equitable world this situation would be resolved.

## Global climate politics

In reality the world reacted throughout the 1990s on a political stage by crafting and creating the United Nations Framework Convention on Climate Change (UNFCCC). Every nation on the planet signed up to the UNFCCC at the Rio Earth Summit in 1992. The objective of the UNFCCC is to bring global greenhouse gas emissions down to a safe level in order that food supply is safe and the atmosphere stabilized. There is still no agreement on what concentration of greenhouse gases in the atmosphere this constitutes, so everyone is assuming that twice the pre-industrial level of $CO_2$ (550 ppm) is the virtual target, beyond which we cross at our peril.

In 1997 the UNFCCC was taken a stage further by the creation of the Kyoto Protocol. This instrument has buried within it various flexible mechanisms (such as Emissions Trading, Joint Implementation and the Clean Development Mechanism) that are designed to bring about a smooth transition of emissions reductions in tandem with technology transfer from the developed to the developing countries. Since December 1997 it has taken four more years to formalize the rules of engagement of the flexible mechanisms, culminating in the Marrakesch accords in November 2001.

Ideally, global emissions have to contract to an end-point (concentration level of, say, 550 ppm) and converge by a given date (say 2050). This would result in global emissions dropping to a more sustainable level of 2–3 gigatonnes (Gt) C/yr as opposed to the present 6 Gt-C/yr or the projected 30 Gt-C/yr (Gt-C/yr is equivalent to 1 billion tonnes of carbon in the form of $CO_2$ emitted annually). This approach is formally known as ‘contraction and convergence’, and was created by Aubrey Meyer of the Global Commons Institute in 1991.

The Kyoto Protocol is clearly not enough. If present emission reductions are delivered this will result in only a 1 per cent cut in global emissions. Set alongside the IPCC recommendation of 60 per cent cut, this is indeed a poor effort. We would need a raft of Kyoto Protocols to deliver such a large cut. If time is of the essence, and political agreement is not there, we have to think hard about the options.

## Renewable energy and carbon management

One practical way to deliver the required greenhouse gas emissions involves a transition from a fossil-based economy to that of a renewable one. Renewable energy in the form of wind, solar, geothermal and tidal are obvious solutions, while the nuclear option is less favourable, despite being carbon-neutral. In reality, a transition from the heavy fossil dependency (presently 80–90 per cent of power is generated by fossil fuels) to a low carbon society will be painful but necessary. The ultimate end point is a carbon-free world, centred on a solar-hydrogen fuel-based economy.

Individual corporations have started assessing their own carbon-based emissions, using the methodology espoused by the Global Reporting Initiative (GRI) approach, with a view to delivering Kyoto-type reductions. Companies like Shell and BP Amoco are conducting internal emissions trading regimes that will inevitably feed into the UK emissions trading regime.

The insurance sector does not directly emit greenhouse gases (unlike the oil sector) and as such is not an immediate target for corporate activity by environmentalists. However, on closer inspection the insurance sector is indirectly responsible for a vast amount of greenhouse gas emissions by virtue of insurance companies' ownership of share equities in most major blue-chip companies that all emit greenhouse gases. Ironically, the insurers actually own the oil companies!

Interestingly, a recent survey of the top fifteen UK insurance companies by Friends of the Earth (2000) showed that very few invested ethically or environmentally. A majority of their collateral is tied up in oil, gas and chemical companies, which collectively are the greatest greenhouse gas emitters. If insurance companies are to be taken seriously, especially if they have signed up to the UNEP-FI statement (2001), they need to realize that their investments will be scrutinized evermore closely in the future. The UNEP-FI, or United Nations Environment Programme–Financial Initiative was created in 1995 as a vehicle to allow financial institutions like banks and insurers to follow a more sustainable path in terms of investments and business practices. Some 100 insurers and 195 banks have signed the initiative to date.

## The financial sector

The financial sector is principally composed of the insurance, banks and pension companies. Globally, this sector cycles US$4 trillion annually (UNEP–FI 2001), the insurance sector being responsible for US$2.3 trillion and the banking sector US$1.7 trillion. These two sectors will be amongst the first of the industrial sectors to be affected by the impacts of climate change. A schematic showing the structure of the insurance sector is shown in Figure 8.1.

Since the 1960s the number of large catastrophic events concerned with weather-related activity have increased almost exponentially. The economic

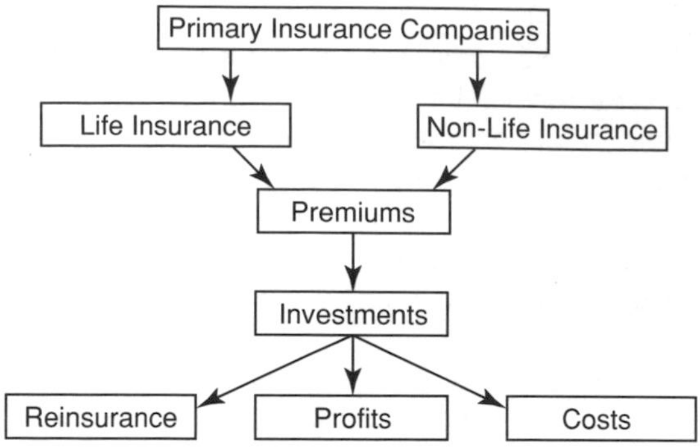


*Figure 8.1* Financial structure of the insurance industry.

and insured losses associated with this increase in activity have grown by almost eightfold and fourteenfold over the last 30 years (see Table 8.1). There are obviously many factors that contribute to these financial losses. Amongst the factors are the following:

- increased wealth;
- increased value of property;
- increased insurance uptake;
- increased weather related hazard.

In terms of large losses in the UK, the real turning points for the insurance industry occurred in 1987 and 1990 when two major wind storms hit the UK (see Homan, Chapter 9 this volume). These events cost £1.4 and £2.1 billion respectively. In the USA, Hurricane Andrew was the ultimate turning point, hitting the eastern seaboard in 1992 and totalling US$16 billion of damage. In the aftermath of these events several US primary insurers, and Lloyd's of London itself, suffered loss of credibility and business. A sea-change in attitudes thus resulted from these events. This could well be a harbinger of future losses.

## Political institutions

Interestingly, at the same time in 1992, the Earth Summit was held in Rio de Janeiro, out of which emerged the United Nations Framework Convention on Climate Change (UNFCCC). The UNFCCC ultimately tasked itself with achieving protection of the climate through stabilization of concentrations of greenhouse gases in the atmosphere to a level that is 'safe'. Five years later, in 1997, the Kyoto Protocol was created with the express aim of bringing about reductions of greenhouse gas emissions by an average of 5.2 per cent for developed countries. Developing countries at present have no emission reduction commitments. This remains a major diplomatic issue in the USA and a barrier to ratification.

The developed countries arrived at a compromise figure of –5.2 per cent by negotiation between the three main power blocks, the European Union (EU), USA and Japan. The UK has the most advanced reduction targets (20 per cent cut by 2010), while the EU are aiming for a 15 per cent cut by 2010 for all three gases combined ($CO_2$, $CH_4$ and $N_2O$). The USA have said that they will never sign the Kyoto Protocol as it stands and would only do so if developing countries take on emissions reductions targets too.

By early 2002, 25 parties have ratified the Kyoto Protocol. The majority of states being either G-77 or small island states, countries with arguably have the most at stake from the future hazard impacts of global climate change. The ratifying states include Antigua and Barbuda, Argentina, Bolivia, Cook Islands, Ecuador, El Salvador, Fiji, Guatemala, Honduras, Maldives, Malta, Mexico, Micronesia, Nauru, Nicaragua, Niue, Panama, Papua New Guinea, Romania, Samoa, Trinidad and Tobago, Turkmenistan, Tuvalu, Uruguay and Uzbekistan (since writing this number has changed to 55 parties. Full details can be found at www.unfccc.de). Embedded within the Protocol are three flexible mechanisms that aim to deliver reductions in ghg emissions by a variety of methods.

1 *Joint implementation* involves the reduction of greenhouse gas emissions between two Annex-1 parties to the UNFCCC (developed country states – essentially the OECD parties) through the use of projects between a donor and receiver state.
2 *Emissions trading* is a trading mechanism to be overseen by the Members of the Parties (a sister designation similar to the Conference of the Parties within the UNFCC structure). This allows member parties to trade emissions credits and permits in order to meet their individual national obligations in terms of emissions reductions targets.
3 *Clean development mechanism* is a vehicle that effectively allows for the transfer of technology from developed countries to developing countries that will directly or indirectly help to reduce emissions in the host country.

The final touches to the institutional structure of the UNFCCC were made in Marrakesh in November 2001. The overall institutional structure of the UNFCCC process has developed over the years, see Figure 8.2.

## Insurance attitudes to climate change

In general, most of the actors in the insurance chain (Figure 8.3) have little or no real perception of climate change issues and how they may impact their businesses; as a consequence they are not taken account of when deciding on business strategy. Indeed, most customers that buy house and car insurance on an annual basis also have a limited idea of how

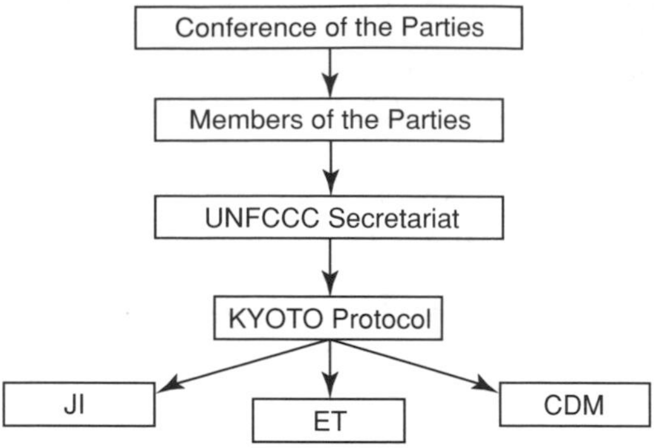


*Figure 8.2* The institutional structure of UNFCC.

Key: CDM = clean development mechanism; ET = emissions trading; JI = joint implementation; UNFCCC = United Nations Framework Convention on Climate Change.

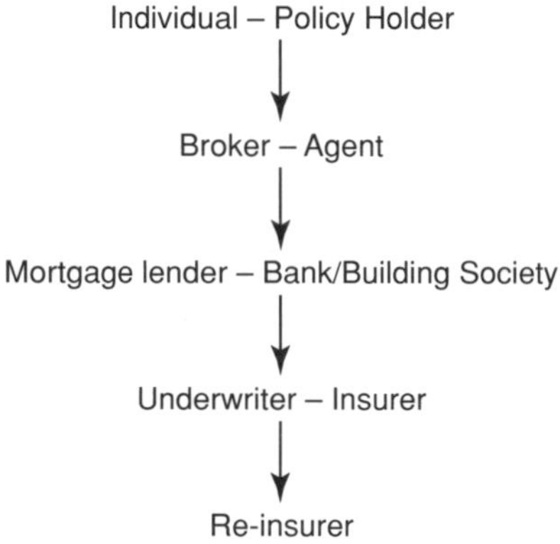


*Figure 8.3* The property risk underwriting chain.

climate change could affect their lives and property. In fact, the perception of weather-related events within the insurance sector has up until now been seen as merely another line of business (similar to airline insurance or life assurance).

As shown in Figure 8.3, insurance is taken out by an individual (e.g. house contents insurance) usually through a broker (on behalf of a primary insurer). Quite often the main insurer is unaware of individual contracts as computer software is used to assess the premium based on past event histories, and as a consequence may be exposed to unknown future risk trends for properties in floodplains or on the coastline. Blocks of insurance can be sold for a group of properties, usually in connection with a mortgage lender – again exposing the insurer to unknown and unquantifiable risks.

Part of the perception problem relates to the fact that each of the actors in the 'insurance chain' operates with a different agenda and associated timescale. The individual policy holder and mortgage lender will be thinking long term (25–30 years), along with the fact that a major financial investment is at stake. The broker, by contrast, considers the transaction as a one-off event. The primary insurer considers her/his exposure as limited to one year, after which time they can get out of the contract. Re-insurers take a longer-term view (of the order of decades) as they ultimately have to absorb the ups and downs of market prices. Re-insurers are ultimately insurers of last resort.

## The costs of climate change

The costs of climate change to the global insurance industry could be huge, even unsustainable. At present the insurance sector cycles US$2.3 trillion annually (UNEP–FI 2001). This represents some 10 per cent of global world trade in value terms.

Typically, the global re-insurance pool at any one time can amount to US$200–300 billion. However, it has recently been stated by United Nations Environment Programme Insurance Industry Initiative (a body created at COP-1 in Berlin in 1995; see www.unep.ch) that potentially the largest climate-related event could amount to US$100 billion in 100 hours. This is based on the worst-case scenario event of a grade-5 hurricane hitting a metropolis such as Miami or New York. This, in contrast, makes the global re-insurance pool look small.

Official UN estimates (IPCC 2001) of the projected cost of climate change (based on projected population and economy growth figures from the UNDP, UN-Population programme) to society amount to 1–2 per cent of world GDP (US$30 trillion). This would make the potential annual cost of climate change impacts amount to US$300–600 billion.

In 1997, Munich Re (2000) estimated that globally the economic costs of weather-related events amounted to US$98 billion, with an associated insured loss of US$12 billion. A further study showed that in the 1990s alone the total economic loss was US$480 billion, with an associated insurance loss of US$107 billion. The full data set is shown in Table 8.1. A simple analysis of these figures shows that something seems to be occurring that should be of concern to the insurance sector. Of greatest concern is the last column, where the costs (economic and insured) have been compared from the 1960s to the present decade. The ratio for the insured losses is a staggering 14:1 compared with the 1960s. Essentially this translates as meaning that insured losses are accelerating at four times the rate of growth of the world economy. Clearly this is unsustainable.

### *Return periods are changing*

If the UK Climate Impact Programme scenarios are to be believed, and there is no good reason to suspect otherwise, then return periods for natural

*Table 8.1* Great weather disasters, 1950–2000

| *Events/losses* | *Decade* | | | | | *Last 10 yrs* | *Factor last 10: 1960s* |
|---|---|---|---|---|---|---|---|
| | *1950–1959* | *1960–1969* | *1970–1979* | *1980–1989* | *1990–1999* | *1991–2000* | |
| Number of events | 20 | 27 | 47 | 63 | 89 | 78 | 2.9 |
| Economic losses (US$)* | 42.6 | 75.7 | 136.1 | 211.3 | 652.3 | 579.9 | 7.7 |
| Insured losses (US$)* | 0 | 7.3 | 12.4 | 26.4 | 123.2 | 103.7 | 14.3 |

Source: Munich Re (2000).

Note: * Losses are in US$ 2001 values.

events such as floods, wind storms and dry hot periods are set to change in frequency in the future. For example, a typically hot summer (like 1995) has a return period at present of 100 years; by 2050 this is expected to have reduced considerably to 1 in 3 (DoE 1996). In insurance terms this means that conventional underwriting of risk (which is based solely on past claims) is somewhat flawed. In the long term this could mean that future claims could exceed reserves, implying that unless climate change impacts are taken into account the insurance sector could become unsustainable.

Evidence for a change in weather patterns may be gleaned from long-term data on such things as El-Niño and La-Niña events. Recent trends suggest that both seem to be occurring more frequently and possibly more severely. There are even theories that the El-Niño/La-Niña oscillation could become more exaggerated by virtue of increased greenhouse gas concentrations. This postulation is conceivable, as increased sea-surface temperatures are being detected every year.

On a more local scale, recent research in Scotland at the University of Dundee (Black and Evans 1999) has shown that over the last 10–20 years a trend of increasing severity and frequency of floods in the west of Scotland is emerging. This is exactly what the UKCIP scenarios have forecast. So, in summary it could be argued that we are already beginning to see the first signs of climate change, even though the insurers are not accounting for it in the underwriting and premium pricing process.

## The UNEP-Insurance Industry Initiative

The United Nations Environment Programme created the Insurance Industry Initiative (UNEP-III) in 1995, coinciding with the first Conference of the Parties (CoP) of the UNFCCC held in Berlin. Its main objective was to create a facility that would allow like-minded insurers and re-insurers to conduct their business dealings in a sustainable manner. An official statement on behalf of the UNEP-III was opened for signature at CoP-1, extolling the virtues of incorporating environmental considerations into financial business activities. To date 100 insurance companies have signed the statement, the companies being mostly from Europe (e.g. Swiss Re) and Japan (e.g. Sumitomo). In addition, six UK-based insurance companies have also signed. These include General Accident (now Aviva), Independent (which ceased to trade in 2001), Iron Trades, NPI, Nat West Insurance Services and Sumitomo (UK). There is an equivalent agreement for the banking sector comprising 195 signatories.

Both initiatives (insurers and banks) have recently been brought together under a new formal arrangement called the UNEP-Financial Initiative (UNEP-FI). Very active work is taking place along three main working groups, involving environmental reporting, climate change and asset management. The climate change working group has sponsored a study into the future of the greenhouse gas market and its implications for insurers. Presentations have been made by UNEP-FI at both CoP-6(ii) in Bonn and CoP-7 in Marakesh. All working groups prepared papers and reports for the World Summit on Sustainable Development (WSSD) in Johannesburg in August 2002.

### *The Tobin tax proposal*

The idea that all financial transactions should be taxed was first mooted by a Canadian economist by the name of Tobin. The incredibly light tax level of 0.01 per cent of the financial transaction would generate revenue that could be pooled into a fund to help promote research and development into mitigation and adaptation methods to help the developed world with the impacts of climate change. If this was applied to the present level of financial flows in the world (US$4 trillion annually) this would generate a fund equivalent to US$400 bn/yr – ironically the level of spend that the UN originally thought would be needed to help fix the climate change problem.

To date this idea has not been adopted, probably through conspiratorial views that this may be the thin end of the globalization and New World Order edge. These discussions, however, may resurface at the World Sustainable Summit on Development (WSSD). In many ways the argument runs alongside the 'Cancel the Debt' campaign run by Jubilee 2000. In fact, the events in New York on 11 September 2001 may add a further voice to this argument as it is beginning to be perceived that underlying much of world tensions at present is the idea that resource depletion, poverty and environmental degradation are the root causes.

## Investments

As mentioned previously, insurance companies own a major share of the equities on the world markets by default. The UK equity market, for example, is valued at around £1.4 trillion. The insurance companies (and pension

*Table 8.2* A possible timetable for decarbonization

| *Date* | *Fossil (%)* | *Renewable (%)* |
|---|---|---|
| 2000 | 100 | 0 |
| 2020 | 80 | 20 |
| 2050 | 50 | 50 |
| 2100 | 0 | 100 |

funds) collectively own between a third to two-thirds of the market at any one time. As a consequence they indirectly own the liability of the greenhouse gas emissions associated with those equities (including oil, coal and gas companies).

If insurers were really proactive and long-term in their investments thinking they should really be actively disinvesting from the fossil-dependent companies in their portfolios and actively investing in sustainable and renewable energy companies of the next century. Such an investment strategy would send a real signal to the markets.

Effectively, society has to be decarbonized over the next few decades to the point that by 2100 (or ideally sooner) all energy is from renewable and sustainable sources. A timescale for such decarbonization could follow the path set out in Table 8.2. The UK government has recently stated that it is targeting a 10 per cent share of the energy supply in Renewable Energy in the UK by 2010.

Once the trend has been initiated, one can expect a rush of new technology companies coming to market with new energy devices and strategies. Solar energy and hydrogen as a fuel are set to be the new versions of oil and coal for the twenty-first century. A UK company, Solar Century (backed by 17 major financial institutions), is attempting to convince the UK government that all new housing should incorporate solar roof panels as a mandatory component of the design.

Major blue-chip companies such as Ford, GM and Lufthansa are looking into the viability of using hydrogen as a fuel for transportation, possibly as a direct fuel or used in conjunction with fuel cells. Ballard, the maker of fuel cells, is enjoying a burst of optimism in its range of products and sees a bright future in the non-fossil fuel sector.

## Policy cover withdrawal

The affordability and availability of insurance cover has always been a central tenet to the UK insurance industry. The UK insurance industry is quite unique in that it offers cover for all forms of peril (wind storm, flood, subsidence, freeze). By contrast many European countries operate a range of policies from 'reserve pools' for certain perils to increasing exclusions on perils such as subsidence, or in some cases total unavailability of insurance for certain areas/perils.

As a consequence it is politically quite difficult for the UK insurance industry to change its cover through the form of policy wording. However, the floods of October–November 2000 (total losses £1.3 billion) made the industry sit up and rethink this position. The Association of British Insurers (ABI) has agreed to hold a moratorium on flood cover until December 2002, under the proviso that the UK government increases spending on UK flood defences. Present annual spending on flood defences is £240 million. The ABI argues that this should be increased by £145 million.

In a world with increased climate hazard risk (and increased weather-related claims) there is a range of policy options for insurers:

1 Increase policy cover premiums.
2 Increase exclusions level on individual claims.
3 Reduce cover.
4 Withdraw cover.
5 Do nothing.

Whichever is adopted will probably be determined by market forces rather than any overriding holistic view of protecting the environment. Each of the above options will have their advantages and disadvantages.

Various companies will take different positions. Some will lead from the front (e.g. Aviva) and take a high moral position, while the majority will follow the mainstream and adapt prices accordingly. The market is beginning to 'harden' (prices increasing) due to a run of bad weather events and the aftermath of the 11 September 2001 event. The re-insurers are back in the driving seat and are steadily driving the cost of re-insurance ever higher. What is really needed is for a general insurance industry consensus on what to do about climate change as a sector. The machinations at the UNEP-FI will hopefully help in this matter.

At present it could be argued that the price of insurance cover is too cheap and does not take into account such issues as climate change impacts. Unfortunately, the insurance industry is 'reactive' rather than 'proactive', tending to wait until an event occurs before doing anything to prevent it occurring in the first place. With climate change there is now enough widespread evidence to inform the insurance industry about potential future impacts. Pricing of the potential damage (of climate change impacts) is the next step that has to be undertaken by the industry, both collectively and individually.

Research into issues such as extreme-value theory and the potential changes in return periods of extreme events such as flood and windstorm need to be conducted by the insurance sector. The full impacts of climate change to the insurance sector need to be assessed. An earlier study conducted by Environmental Resources Management looked at all sectors in the UK. The insurance sector was assessed as being at medium/high risk of impact by climate change. No monetary figure was placed on the future costs of climate-induced damages.

A second study conducted by the Building Research Establishment (2000) attempted to price the individual components of the building sector that needed to change to climate change impacts. In total the potential cost to the UK economy was estimated to be as high as £30 billion.

## Insurance reactions to natural disaster

The active hurricane season of 1992 culminated in Hurricane Andrew hitting Florida and led to total insured losses of US$16 billion and economic losses in excess of US$30 billion. Only 74 people lost their lives in total, yet five US-based insurance companies became insolvent as a result. As a consequence hurricane cover was completely adjusted in the region at risk by the federal government legislating to force insurers to offer mandatory cover, backed up by government pools of reserves.

In 1998 Hurricane Georges hit Puerto Rico, Hispaniola, Cuba and, finally, Florida and Mississippi. A new loss record was set for the region with economic losses reaching US$10 billion. In total 4,000 people died, most of them in the Dominican Republic. 1999 brought Hurricane Hugo, with associated economic losses of US$9 billion and insured losses of US$4.5 billion. The death toll was only 87.

Hurricanes are not the only major events that cause losses. Volcanic explosions such as the recent event on Monserrat are also devastating – probably more so than a hurricane as the damage to the land is long term and generally irreversible. The explosions in Monserrat caused insurers to run for the hills, avoiding paying many of the claims that followed from future explosions. In a similar vein, the recent floods in the UK (1998 and 2000) have caused the UK insurance industry to demand that the UK government spend a lot more heavily on flood defences otherwise the unthinkable act of denying flood cover in parts of the UK could become a reality. Thus if one glimpses the future through a climate-change glass ball one could readily envisage areas of the world that will become uninsurable. For instance, Egypt, the Netherlands, Bangladesh, parts of southern India, areas of East China, and small Pacific islands could all be excluded because of their risk of being inundated by high seas and large storms caused by global warming effects.

### *Planning*

The other major issue that has come to the fore recently is the problem of planning consent, particularly with respect to new buildings in floodplains or on contaminated land. As the world gets warmer there will be a tendency for more flooding to occur both inland and on the coasts. Thus, existing buildings will have to be proofed against floods (anti-flood devices) as well as strengthening conventional flood defences in the catchments and on the coastlines (e.g. sea walls). However, in order for this to have any real effect it has to be backed up by strong planning laws that allow vested interests such as insurers to object to new planning (DEFRA 2001) applications in high-risk areas such as known floodplains. Insurers need to sit on the planning application boards when such cases go through the official channels, otherwise they could be lumbered with massive flood-related payouts that could otherwise have been avoided.

### *Insurance models for the future*

If insured and economic losses continue to spiral at the present-day rates (+10 per cent/yr) as against a 2–3 per cent growth in wealth of the global economy, then the financial system as we know it will not be able to hold the strain. The present insurance model will thus have to be adjusted to a new and more adaptable form. Already the emergence of 'weather derivatives' is a move in this direction, in effect transferring the risk of loss to the futures market. These products were devised as a new vehicle to do just this in reaction to the potentially large losses that appeared to be emerging from natural events cover.

Some countries, such as the Netherlands and France, already operate a national reserve pool to cover the excess losses that cannot be covered by individual insurers. The recent storms in 1999 have virtually depleted the French reserves, requiring everyone to pay an extra 10 per cent on the premiums for the following year to top up the pool.

Ideally, for an issue that is as global as climate change a global solution is required. The Tobin tax idea comes close to this. As explained earlier, such a tax on financial transactions would raise US$400 bn/yr – which would go some of the way to paying for the mounting losses. The 1990s saw a total loss (economic and insured) of some US$700 billion and will soon tip the US$1 trillion level. The irony, of course, is that the majority of losses will occur in the developing countries and will ostensibly have been indirectly caused by the emissions (of greenhouse gases) from the developed countries. Thus, one could argue the case for a massive transfer of wealth from the rich to the poor (in the form of insurance cover) to offset the damage that has and will be dealt to the developing world by the richer nations of the world.

Another, less direct way in which insurers could help alleviate this potential problem of crippling future losses would be to invest in alternative energy technologies. Others are copying the trend, although the actual amount invested in these funds is trivial compared to the major markets. A recent UNEP study has identified over thirty-five such funds.

Insurers could demand that planners adopt improved building standards and regulations, especially for properties being built in the high risk zones of the future climate world. If properties are not built to the more rigorous demands of future climate impacts, insurers would be entitled to pull cover for these regions. Pressure could thus be brought to bear on the various layers of the property chain to enforce these required changes.

## Conclusion

Climate change is happening. The rate of change is unknown, but the direction is more certain. The impacts of climate change may occur more severely and more frequently than otherwise considered and as a consequence will cause a lot of financial woe for the insurers (and indirectly pension holders). It is thus imperative that insurers reassess their present method of underwriting in the face of a known and complex risk that is quite difficult to quantify. They need to take proactive action on many fronts, including policy wording and investments. If they do not take action the ultimate irony may be that the insurance industry might suffer from its own inward-looking practices. It has to forget the present practice of thinking of short-term gains and operating on an annual cycle and adjust to a long-gain strategy with a more holistic global viewpoint. The insurance industry has an important corporate role to play in today's society by virtue of its immense leverage in the markets. If it chooses to ignore these early warning signs there may be no insurance industry to speak of in 20 years!

## References


Black, A. and Evans, S. (1999) *Flood Damage in the UK: New Insights for the Insurance Industry*, Dundee: University of Dundee Press.

Building Research Establishment (2000) *Implications of Climate Change Impacts on the Construction Sector*, Watford, Building Research Establishment.

Department of the Environment (DoE) (1996) *Hot Summers Report*, London: DoE.

Department for Environment, Food and Rural Affairs (DEFRA) (2001) *Policy Planning Guide, PPG25*, London: DEFRA.

Friends of the Earth (2000) *Capital Punishment*, London: Friends of the Earth.

Intergovernmental Panel on Climate Change (IPCC) (2001) *Third Assessment Report*, Cambridge: Cambridge University Press.

Munich Re (2000) *Topics 2000*, Munich: Munich Re.

United Nations Environment Programme – Financial Initiative (UNEP–FI) (2001) *United Nations Environment Programme – Financial Initiative Position Statement*, CoP-7, Marakesh.

# Part IV

# Local contexts and global pressures

# 9 The social construction of natural disaster

## Egypt and the UK

*Jacqueline Homan*

> there are no expert solutions in risk discourse, because experts can only supply factual information and are never able to assess which solutions are culturally acceptable.
>
> (Beck 1996: 20)

## Introduction

Disasters researchers engaging with broad post-structural work on society–nature relations are beginning to recognize the numerous factors that impinge on social constructions of disaster, influencing the ways in which people perceive and respond to risk and provoking much local variation both within cultures and societies and between them (e.g. Blaikie *et al.* 1994). This chapter explores natural disaster perception through two case studies: an earthquake that struck Cairo, Egypt on 12 October 1992 and storms that affected the south coast of the UK on 16 October 1987. These cases reveal that although recent thinking in sociology and human geography has indicated that globalization of culture is occurring (e.g. Harvey 1990; Hannerz 1990), environmental meaning remains culturally and ethnically constructed and subject to the idiosyncrasies of indigenous culture.

## The social construction of disaster

### *Understanding the 'natural' in disasters*

Whilst contemporary research indicates that exposure to disaster is far from 'natural' and that power inequities often explain differences in losses sustained and recovery, perception studies (including the case studies used in this chapter) reveal that people tend to view events as an exaggeration of the natural world. It is therefore important to explore the ways in which social constructions of nature have changed within societies in order to contextualize hazard perception.

Research into 'natural' disasters and risk has, until relatively recently, been typified by modernist approaches and confined within expert discourses. As such, it has neglected to account fully for cultural idiosyncrasy and social constructions of both the natural world and the hazards that occur within this world. Indeed, hazards research has tended to separate society and nature, with hazards being seen as external to the everyday and mitigation becoming the preserve of an expert elite.

Debate within social science has increasingly tried to reconcile and find meaning behind the society–nature dualisms that pervade contemporary society (e.g. Dickens 1992, 1996) and has given considerable attention to exploring the social construction of nature (e.g. Eder 1996; Macnaghten and Urry 1998; Phillips and Mighall 2000; Castree and Braun 2001). This research has identified that perceptions of nature differ over time and in place (Simmons 1993; Urry and Macnaghten 1995; Demeritt 2001). The way in which natural hazards are understood and, as a result, mitigated, should be informed by this evolving understanding of the natural world. For example, rather than the modernist view of nature and natural hazards as being phenomena that can be dominated and 'tamed' (e.g. through hard engineering approaches), the development of a more symbiotic relationship between people and the hazards with which they live (e.g. though floodplain planning) could result in less vulnerability.

### *Trust me, I'm an expert: power in the risk society*

With the changing perception of nature, the view of human agency has also changed from one of domination, whereby experts were in control of knowledge, to a more horizontal relationship where there is equal recognition of lay knowledge. With this has come acknowledgement of power imbalances – for example, between economically strong and weak states and between experts and lay people. In particular it has been observed that dominant discourses of science and engineering have historically been controlled and shaped by a white, western, male elite (e.g. The Ecologist 1993; Escobar 1996).

A particularly pertinent aspect of this power imbalance is (mis)trust; experts have typically withheld information, as there is a preconception that the public will not understand it, and the public do not trust experts to tell them the truth. The former instance has been described as the public living with a false consciousness, that they are unaware of their 'real' situation, or that they are in a position of ignorance (the deficit model), whereas experts alone are aware of the 'truth' (e.g. Millar & Wynne 1988; Wynne 1991; Locke 1999). Whilst it is the case, as acknowledged above, that experts and lay people are likely to construct a risk issue differently, this does not mean that one kind of knowledge is essentially more or less valid.

The persistence of expertise-driven natural disaster research has resulted in a deskilling of lay people; knowledge has passed into the hands of a few and has become more specialized. Increasing losses as a result of natural hazards (e.g. IFRC/RC 1999; Chester *et al.* 2001) have partly been as a result of structural weaknesses, such as the economic situation and social conditions, but are also related to the social distribution of knowledge (Zaman 1999).

***The 'cultural turn' in hazards research***

Two paths have emerged in hazards research – one ascribes a central place for scientific approaches, and another that places people at the centre of mitigation (e.g. Chester 1993; Oliver-Smith and Hoffman 1999). Both approaches have tended to proceed from the classical modernist view that there are 'real' ways of looking at hazards which correspond to a truth about the world, and that nature is seen as separate from society. As a result, whilst the practical act of mitigating hazards has become more culturally sensitive, the theoretical discourse supporting these views remains wanting.

The pervasiveness of the modernist view manifests through a tradition of domination by scientific discourse and, until recently, a tendency to ignore the world that exists beyond rational thought. Metaphysical understandings, for example, were, for a long time, seen as being irrational and superstitious ways for people to make sense of the natural world and the hazards occurring within it (e.g. Hewitt 1983). Further, the western science adopted in hazards research constructed the global North as being rational and progressive with the global South regarded as irrational and backward. As Rabinow (1986) notes, the West also now needs to be anthropologized, to

> show how exotic its constitution of reality has been; emphasize those domains most taken for granted as universal; . . . make them seem as historically peculiar as possible; show how their claims to truth are linked to social practices and have hence become effective forces in the social world.
>
> (Rabinow 1986: 241)

Acknowledgement of the value of indigenous, non-expert constructions of reality demonstrates a concession to the social constructionists. To date, research has primarily focused on perceptions in Latin America, Africa and Asia, although it is often the case that an indiscriminate use of overtly technocratic solutions has also dominated mitigation in the North without acknowledgement of cultural variations and adaptations (e.g. Bolin and Stanford 1998). However, such technology is androcentric and ethnocentric, and therefore many, even in the West, are alienated from both its development and application.

## Case studies

***Egypt: the socio-cultural contextualization of seismicity***

Egypt is a country with an extensive and well-documented record with regard to earthquakes (e.g. Ambraseys *et al.* 1994). However, when a moderate-sized event occurred 35 km to the south of Cairo (at Dahshûr) on 12 October 1992 it caused a disproportionate amount of damage. The magnitude of the

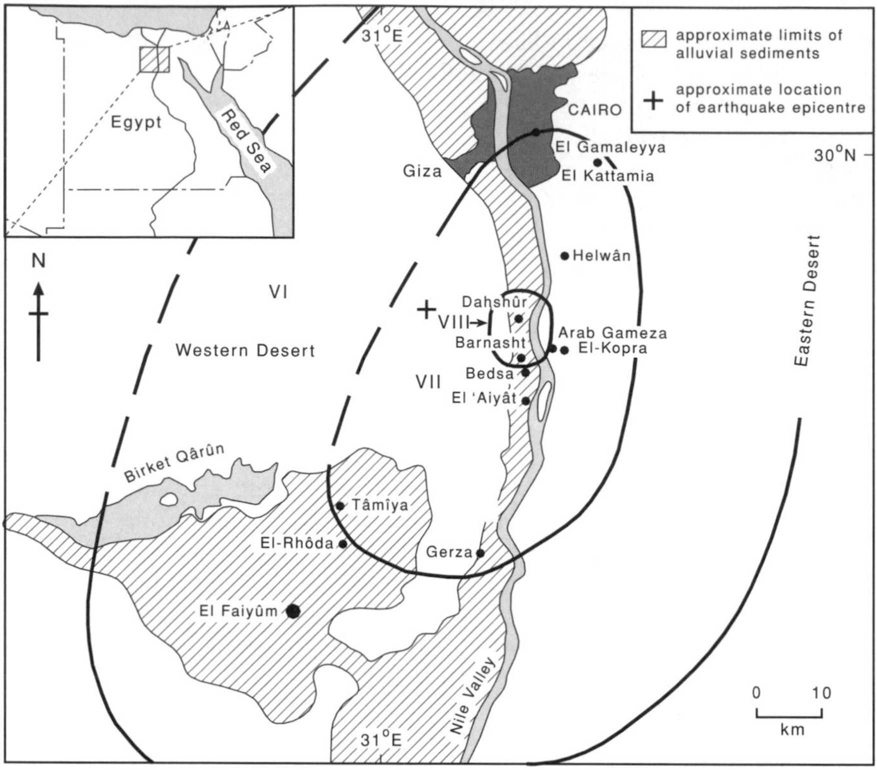


*Figure 9.1* A location and isoseismal map of the 12 October 1992 Dahshûr, Egypt earthquake [Adapted from Degg 1993).

1992 event was 5.6, according to the Helwan Seismic Observatory, and intensity, according to the Modified Mercalli intensity scale, peaked at VIII. Approximately 551 people were killed in this earthquake and almost 10,000 were injured. The material presented below is derived from interviews with community members conducted by the author in Gerza, Bedsa, Barnasht, El Kattamia, El Gamaleyya and central Cairo (Figure 9.1).

## *The nature of knowledge and the knowledge of nature: Islam and the hazardous natural world*

To understand earthquake perception within an Egyptian context it is essential to understand the importance of the Koran, which is viewed as a guide to life in both an everyday and more metaphysical way. Perception of nature is linked with religious interpretation – for example, concepts of 'oneness' between God and nature; that humans are stewards of the Earth but that ultimately control is maintained by other-worldly forces; and, most fundamentally with regard to the present discussion, that nature is the means by which God demonstrates His superiority over humans.

With regard to the latter point, the Koran is comprised of a number of *surahs* (chapters) that link natural hazards with pecuniary actions and thus make explicit links between this-worldly forces and other-worldly causes. Examples of the kinds of hazards described are famine and earthquakes, the latter comprising the subject matter of an entire *surah* (called 'The Earthquake'), which clearly states that Judgement Day will be denoted by the occurrence of an earthquake.

Having identified religion as a central defining concept in Islamic interpretations of nature, it would be a mistake to assume that construction of nature is based solely on the Koran. Science also has a key role and is combined in a complex way with religious understanding, in particular through the way that knowledge is pursued. It is not Islamic to pursue knowledge for its own sake, but rather to develop a gnostic and esoteric understanding of the world. For example, Negus (1992) outlines two kinds of science:

> One is based on nothing else but human thought and the belief that the physical universe is the only reality. It is basically atheistic and is outside of Islam. The science that is *within* Islam differs from this because it always starts with the view of the world given in the Quran.
>
> (Negus 1992: 39)

Whilst some have maintained that there is evidence for every theory and event in the Koran (e.g. see Akhtar 1990), on the whole the linkages between science and religion are seen in a much more flexible way; for example, Dhaouadi (1992: 198) maintains that 'the Islamic base of science and knowledge is *multi-dimensional*'. There is therefore freedom to develop an eclectic and flexible approach towards knowledge, and for diversity and heterogeneity within the overarching discourses provided by the Islamic world-view. Therefore, while other-worldly considerations may remain the umbrella under which this-worldly pursuits take place they are not a constraining influence. Indeed, religious interpretation is usually the way that people try to understand *why* such devastation occurs as opposed to a pragmatic strategy to deal with disaster (e.g. Schmuck 2000). However, despite the central importance of these issues there has been a conspicuous lack of discussion on the linkages between theology and natural disasters (Chester 1998).

### *Seismic culture? Earthquake perception in Egypt*

Of the 136 people interviewed in Egypt, the majority (116) felt that earthquakes were either wholly, or in part, of direct religious origin; many drew on apocalyptic imagery, thus linking their understanding of earthquake cause with the Koran in a very literal way:

> There was banging during the earthquake. As well as the banging the sky turned red and there were flashes across the sky; we thought of it as the Day of Judgment.
>
> (Bedouin sheikh, village of Bedsa)

> The earthquake was related to God. It was sent as a memorandum from God to show how strong He is in both place and time.
>
> (Security man, aged 40, village of Gerza)

> It [the earthquake] was sent due to the anger of God because of the behaviour of the people.
>
> (Female shopkeeper, aged 47, interviewed in El Gamaleyya)

Also important is the interpretation of earthquakes as scientific events within the overarching discourse of religion; that is, that whilst this-worldly explanations have some validity they only do so within the context of Islam:

> I was asleep when the 1992 earthquake happened. I was woken up and I thought that it was something sent by God. I went downstairs and saw people shouting and crying and looking for their families. Then I knew it was an earthquake and I asked God to make it stop before there was any more damage . . .
>
> Everything comes from God, only He can save our souls, *but* you can also explain earthquakes scientifically . . . There are scientific controls on the environment and earthquakes and so on are natural events, but they are ultimately controlled by God. You can have science but this comes from God.
>
> (Male university graduate, aged 35 interviewed in Cairo)

Other interpretations of hazard are metaphorical in nature, perhaps emanating from folklore or historical association and, indeed, Sayer (1992) stresses the importance of explaining the unfamiliar through 'picture-carrying expressions'. In the case of earthquakes, which are likely to be encountered only infrequently, such linkages to familiar objects or events as a way of explanation are inevitable. In Egypt, for example, ancient Egyptian building practices were said to have made the ground unstable and, in one case, the following analogy was described:

> There is a theory amongst some people in Egypt that the earth is balanced on cow's horns. The cow holds the earth on one horn and then, to be more comfortable, will move it across to the other horn. This movement is what causes earthquakes.
>
> (Respondent as in previous interview extract)

It is also the case that many of those who sought scientific explanation did so through visualizing what might have happened in the earth's crust to trigger an earthquake. Amongst phenomena mentioned were excessive heating from the earth's interior; a build up of water pressure; and the notion that there is an 'earthquake' belt that countries physically pass through.

Finally, in the case of Egypt, there were also a small number of people (five) who did not really know why the earthquake had occurred, just that seismic events are dominant in different parts of Egypt or the world as a whole. Interestingly, of those who did not speculate as to a reason for the earthquake all were living in the communities that had been less affected by the event, possibly reinforcing the concept that the greatest need to understand comes from those that have experienced the greatest losses.

### *Practical action: responding to seismicity*

Although numerous people described the earthquake as caused by otherworldly forces, it would be wrong to assume that this results in a fatalistic attitude towards disaster. Religious explanation is meaningful to people but very few felt that this precluded practical action. For example, the following indicate that people actively sought further information as a consequence of the earthquake as to what to do in a future event:

> I have trained myself to face any other earthquakes and I will know how to behave in the future. I got information about this from newspapers, radio, TV and magazines, and I trained myself in the aftershocks. I have learnt that I should stay at home under a table/in a doorway.
>
> (Unemployed male teacher, village of Gerza, aged 25)

> If an earthquake happened again, people would know how to behave. This is unlike the last earthquake where a lot of children died, not due to building collapse but because of suffocation in their panic to try and get out of the building. A book was produced to tell people what to do if an earthquake happens again – for example, they should stay under tables, turn off the lights and stay calm.
>
> (Male hotel worker, aged 30, interviewed in Cairo)

Within Egypt, the strongest link to religion and practical action presents itself in issues surrounding gender; an interview with a young woman in Cairo indicated that in the 1992 earthquake some women were placed at greater risk than men because they were reluctant to leave their homes with their heads uncovered and hence took a longer time to leave buildings in danger.

### *The anthropology of disaster in the West: natural hazard perception in the UK*

Considerable attention has been given to redressing many of the inappropriate approaches to disaster in the global South, whilst an indiscriminate use of technological solutions have continued to be applied within the West. This section explores perception of a meteorological hazard event in the UK and, in the light of this, indicates potential problems of an over-reliance on technology.

On 16 October 1987, a depression developed in the Bay of Biscay and then progressed across the south coast of the UK. It struck the south coast of Cornwall at 23.00 with winds of 110 m.p.h. and eventually reached London at 07.00 with winds of 94 m.p.h. The storms were classed as 11 (violent storm force) on the Beaufort scale (*The Times* 1987). The damage caused was primarily to infrastructure, although 17 people were killed in the event (Randall and Bowcott 1987).

#### *Changing interpretations of nature in the UK: the move to a plural world-view?*

Understandings of the natural world in the UK have gone through numerous transitions, although three main phases can be identified: pre-modernism, modernism and post-modernism. On a very generic level, pre-modernism was constituted by an holistic interpretation of the natural world, with religious discourses being seen as particularly dominant; scientific theories to explain the natural world in ways which conflicted with religious doctrine were met with either scepticism or hostility. With the Industrial Revolution, modernity began to privilege rationality and reason over religious belief; scientific and religious interpretations of nature were seen as mutually exclusive (Habermas 1990). Interpretations of the world, and of nature, became increasingly secularized (e.g. Berger 1969, 1981; Berger *et al.* 1973; Weber 1992; Gellner 1992; Bruce 1996) and with this change in perception came a change in intervention; nature became viewed as an external object and something to be dominated (e.g. Glacken 1967, Simmons 1993; Harvey 1996). Risk, as opposed to being other-worldly, became increasingly associated with this-worldly realms. As a result expectations rose of the capabilities of people (experts), to solve the problems of risk and hazard (e.g. Bauman 1987, 1988, 1992; Giddens 1990, 1991; Beck 1992, 1996).

In the West the advent of postmodernity in the 1980s onwards prompted change towards a more eclectic perception of reality. Alongside this came an awareness of the limitations of scientific explanation and the role of an expert elite. Where once there had been optimism there is now increasing pessimism and a romanticization of nature (e.g. Gandy 1996), accompanied by the rise of popular culture and increasing recognition of the credibility of lay perspectives has become significant. New Age philosophies have become increasingly popular and, with this, a more eclectic and inclusive environmental discourse has developed (e.g. Douglas 1988; Bruce 1996).

From the unproblematic acceptance of hazards as Acts of God through to the more profane understandings offered by discourses of modernity, there is now the possibility for more eclectic interpretations where people are able to construct a world-view from perspectives that were once seen as conflicting (e.g. science and religion). This can result in convoluted interpretations of the hazardous natural world, which present numerous challenges within hazards research if we are to fully incorporate social and cultural perspectives into practical initiatives.

### *Natural hazard in the UK: a case study from the 1987 storms in the South-east*

The case study used as an example of 'natural' hazard perception in the UK is based on autobiographical accounts held in the Mass-Observation Archive at the University of Sussex in the UK. The Archive, set up in 1937, holds qualitative accounts on topical issues collected from volunteers amongst the general public. The responses people give contain rich anecdotal and factual material. Following the severe storms that affected the south coast of the UK on 16 October 1987 volunteers were asked to recollect their personal experiences and feelings of the event; coping strategies and practical issues surrounding the aftermath; any changes in the way people feel that they might live following the event; and views on government action that should have been taken as a result of the event. The issues covered thus reflect both perceptual and practical issues, with the two being closely linked.

#### PERCEPTIONS OF THE 'HURRICANE'

The 'hurricane' in the south of England led some people to see God as being in control of the elements:

> 'Weather forecasters are mere mortals'. The creation of big winds and the direction in which they blow is still to a degree in the hands of THE ALMIGHTY . . . Perhaps when touching on the subject of weather . . . you would think that people would be reminded we are still mere specks of dust in relation to the events of the Universe.
>
> (Male, aged 61)

A further interpretation is that the 'hurricane' was sent by God to remind people of the hierarchical relationship that exists between this-worldly and other-worldly forces:

> 'From time to time GOD SHOWS THAT HE IS MASTER'. The theme of our Parish Priest's Homily the Sunday after the gales. 'Man may get to the

moon, do wonderful things with technology, think he can have all the material things of this life, BUT just now and again He does something like that just to prove that He is Master of all.'

(Female, aged 65)

The 'hurricane' of 1987 led to a conception of nature as being in control:

> Nature had gone mad, out of bounds – destructive, irresponsible, racing to its death, dragging everything with it. This was it. The end of the world was here – now, at this moment, and nothing could stop it. Breaking glass, falling and breaking objects and the all-powerful noise that would not tolerate anything above it or mitigate it, took over my brain . . . All reason and life had stopped as we know it.
>
> (Female, aged 48)

*Disaster prevention: blaming the experts*

In assessments of practical response, discussion focused on the function and usefulness of expertise. The following is representative of those who came down in defence of experts:

> There was strong criticism of the Met. Service and they make useful whipping boys, though no one seems to know what action could have been taken if there had been more warning . . . No matter how much money you throw at it, the problem is formidable and our Met. office does a fine job.
>
> (Male, aged 69)

In contrast to this are examples of those who sought to blame someone or something for the devastation; indicative of the increasing disillusionment with respect to science, this was often the expert. For example:

> the weathermen GOT AWAY WITH MURDER . . . Some say what good would it have done to know in advance? Well, firstly they SHOULD HAVE KNOWN IN ADVANCE since they push constantly all the codswallop about mod cons, satelites [*sic*] . . . space stations and so forth . . . People seem to forget that SOME PEOPLE ACTUALLY *DIED*! . . . all brushed under the carpet as usual to suit those who did not want to answer for it or take responsibility.
>
> (Female, aged 56)

Finally, some see the need for experts to listen to laypeople. This tends towards a more post-modern conception of the world, with a levelling between expert and laypeople:

I think the weathermen, and other powers that be, should at least listen to ordinary people when they issue warnings of this sort.

(Female, aged 40)

### *Egypt and the UK: same or different? The importance of context*

The material collected indicates that whilst perception can be similar in instances of environmental disaster – for example, as the two case studies demonstrate, people in different places look for other-worldly causes to explain the devastation and provide a purpose for the suffering – the context in which these perceptions are situated are important. In the UK, religious explanation predominantly comes from those who believe that it offers a more valid explanation than expert discourses, particularly in light of the current disillusionment with science. In contrast, in Egypt other-worldly causes are seen as overriding all explanation not only of disasters but also of the natural world more generally, and thus discourses such as science are contained within this larger discourse. Therefore, whilst religious explanation is given in both instances there is different motivation as to why; it is therefore essential that context, such as that between people and nature, is understood.

The reliance on religious explanation indicated by a large number of respondents also indicates the importance of understanding this aspect of disasters research more clearly. Complex interactions between nature and religion and, in particular, between religion and disaster (e.g. discussions on theodicy) have tended to be lacking; as Chester (1998) notes, this needs to be redressed. Such understandings should aid the natural disasters researcher in their understanding of perception, and thus social and cultural construction of disaster (e.g. Mitchell 2000).

The different approaches towards science and expertise in the two societies reinforces what is already known within disasters research – that motivation for, and the type of, mitigation needs to come from the grassroots (e.g. Maskrey 1989, 1995; Coburn and Spence 1992). It is increasingly the case within development studies that there is a move away from external forces (e.g. from Western countries) driving projects, and some have suggested that development itself is an ethnocentric concept that is imbued with power implications and thus should be avoided altogether (e.g. Escobar 1992). Perceptions from Egyptian interviewees would indicate that practical measures are more likely to be adopted if they are presented in an other-worldly context, and thus are inevitably culturally constrained. Indeed, following the 1992 earthquake the Egyptian government produced a booklet entitled *Earthquakes, Catastrophes and the Role of People in Facing Them* which presented science and practical, hazard-reducing measures in an Islamic context (Homan 2001). However, it is also clear that in the UK there needs to be a move away from expert-led approaches and further involvement of laypeople in natural disaster mitigation. Indeed, the Environment Agency of

England and Wales is increasingly trying to move away from an expert-led culture and enter into more deliberative environmental decision-making processes with the public, for example through its project exploring participatory risk assessment (Petts *et al.* forthcoming). However, what it is also important to note is that the 'rational' approaches that were typically seen as characterizing Western society are to some degree a myth – indeed, there are a plethora of varied and increasingly convoluted explanations of the natural world and disasters. This is likely to increase with post-modernity and the growing access people have to 'other' belief systems and world-views – for example, the increase in Eastern religion in the West.

A further parameter worthy of note at this juncture, particularly in terms of contextualizing global risk society, is that of manufactured risk. These 'human-induced' risks may result in increasing polarity between the 'West and the rest' with regard to disaster perception as a whole, and certainly with regard to the significance ascribed to 'natural' risks (Giddens 1998; Pelling 2001). In particular, the evolution of genotechnology and food risks, such as BSE for example, has given many in Western societies a relativity in terms of risk perception that was previously lacking; that is, there is less concern about what nature can do to 'us' and more concern about the results arising from what 'we' do to nature (Giddens 1998). In contrast, many societies in Africa, Asia and Latin America are still dealing with natural disasters such as earthquakes, floods and plagues, but are also having to cope with the newer manufactured risks that result from imported and transferred technologies (Pelling 2001). Global society is therefore facing new challenges, but the policy changes that accompany this are predominantly confined to the West. For example, in the UK there has been much attention given to the issue of science and policy (e.g. House of Lords 2000) and the need to regain public trust as part of the democratic process through wider engagement in scientific debate. However, in addition to these more policy-oriented issues we are also faced with theoretical challenges – for example, what do we now mean by a 'natural' disaster? Are some disasters more natural than others? How will differential abilities to cope affect pre-existing global inequalities with regard to risk?

## Conclusions

Approaches to nature, science, and disaster have changed considerably over time; there is now recognition of a greater plurality of understanding and a demise of the singular world-view that was the preserve of a modern expert elite. As part of this changing approach, social constructionism has made an important contribution through fostering an acceptance of the validity of all forms of knowledge.

This chapter has illustrated the importance of understanding the contextual 'backdrop' to disaster occurrence and the deep-rooted beliefs that impinge upon perception of the natural world. In particular it suggests a post-structural approach towards the exploration of natural disasters; that is, a questioning of the rigid categorization of disasters and their perception and a move towards a more fluid understanding of 'meanings and identities, forces and relations' (Braun and Wainwright 2001: 60). Adopting a more flexible and reflective approach that post-structuralism enables facilitates a depth of understanding that allows for exploration of the idiosyncrasies and subtle changes in society that influence disaster perception. For example, whilst on a superficial level it might be observed that people in Egypt and the UK will try to cope pragmatically with the hazards and disasters that impinge upon their lives irrespective of social and cultural constructions of the events, the rationale for acting in a particular way is part of a much wider context. Indeed, it is only by understanding the deep-rooted beliefs and causal mechanisms that influence the ways in which people explain the hazardous natural world that culturally acceptable solutions to disaster that are meaningful and acceptable to people are likely to be adopted.

## Acknowledgements

The Mass-Observation Archive material has been reproduced with permission of Curtis Brown Group Ltd, London, on behalf of the Trustees of the Mass-Observation Archive Copyright Trustees of the Mass-Observation Archive.

## References

Akhtar, S. (1990) *A Faith For All Seasons*, London: Bellew.
Ambraseys, N.N., Melville, C.P. and Adams, R.D. (1994) *Seismicity of Egypt, Arabia and the Red Sea: A Historical Review*, Cambridge: Cambridge University Press.
Bauman, Z. (1987) *Legislators and Interpreters: On Modernity, Post-modernity and Intellectuals*, Cambridge: Polity Press.
Bauman, Z. (1988) 'Viewpoint: sociology and postmodernity', *The Sociological Review* 36 (4), 790–813.
Bauman, Z. (1992) *Intimations of Postmodernity*, London: Routledge.
Beck, U. (1992) *Risk Society: Towards a New Modernity*, London: Sage.
Beck, U. (1996) 'World risk society as cosmopolitan society? Ecological questions in a framework of manufactured uncertainties', *Theory, Culture and Society* 13 (4), 1–32.
Berger, P.L. (1969) *The Social Reality of Religion*, London: Faber.
Berger, P.L. (1981) *Modernisation and Religion*, New York: Brunswick Press Ltd.
Berger, P.L., Berger, B. and Kellner, H. (1973) *The Homeless Mind*, Harmondsworth: Penguin.
Blaikie, P., Cannon, T., Davis, I. and Wisner, B. (1994) *At Risk: Natural Hazards, People's Vulnerability and Disasters*, London: Routledge.
Bolin, R. and Stanford, L. (1998) *The Northridge Earthquake: Vulnerability and Disaster*, London: Routledge.

Braun, B. and Wainwright, J. (2001) 'Nature, poststructuralism and politics', in N. Castree and B. Braun (eds) *Social Nature: Theory, Practice and Politics*, Oxford: Blackwell.

Bruce, S. (1996) *Religion in the Modern World: From Cathedrals to Cults*, Oxford: Oxford University Press.

Castree, N. and Braun, B. (eds) (2001) *Social Nature: Theory, Practice and Politics*, Oxford: Blackwell.

Chester, D.K. (1993) *Volcanoes and Society*, London: Edward Arnold.

Chester, D.K. (1998) 'The theodicy of natural disasters', *Scottish Journal of Theology* 51 (4), 485–505.

Chester, D.K., Degg, M., Duncan, A.M. and Guest, J.E. (2001) 'The increasing exposure of cities to the effects of volcanic eruptions: a global survey', *Environmental Hazards* 2, 89–103.

Coburn, A. and Spence, R. (1992) *Earthquake Protection*, Chichester: Wiley and Sons.

Demeritt, D. (2001) 'Being constructive about nature', in N. Castree and B. Braun (eds) *Social Nature: Theory, Practice and Politics*, Oxford: Blackwell.

Dhaoudi, M. (1992) 'An operational analysis of the phenomenon of the other underdevelopment in the Arab World and in the Third World', in M. Albrow and E. King (eds) *Globalization, Knowledge and Society*, London: Sage.

Dickens, P. (1992) *Society and Nature: Towards a Green Social Theory*, Philadelphia: Temple University Press.

Dickens, P. (1996) *Reconstructing Nature: Alienation, Emancipation and the Division of Labour*, London: Routledge.

Douglas, M. (1988) 'The effects of modernization on religious change', *Daedalus* 117 (3), 457–484.

Ecologist, The (1993) *Whose Common Future? Reclaiming The Commons*, London: Earthscan.

Eder, K. (1996) *The Social Construction of Nature*, London: Sage.

Escobar, A. (1992) 'Development planning', in S. Corbridge (ed.) (1995) *Development Studies: A Reader*, London: Arnold.

Escobar, A. (1996) 'Constructing nature: elements for a poststructural political ecology', in R. Peet and M. Watts (eds) *Liberation Ecologies: Environment, Development, Social Movements*, London: Routledge.

Gandy, M. (1996) 'Crumbling land: the postmodernity debate and the analysis of environmental problems', *Progress in Human Geography* 20 (1), 23–40.

Gellner, E. (1992) *Postmodernism, Reason and Religion*, London: Routledge.

Giddens, A. (1990) *The Consequences of Modernity*, Cambridge: Polity.

Giddens, A. (1991) *Modernity and Self-Identity*, Cambridge: Polity.

Giddens, A. (1998) 'Risk society: the context of British politics', in J. Franklin (ed.) *The Politics of Risk Society*, Cambridge: Polity.

Glacken, C.J. (1967) *Traces on the Rhodian Shore: Nature and Culture in Western Thought from Ancient Times to the End of the Eighteenth Century*, Berkeley: University of California Press.

Habermas, J. (1990) *Moral Consciousness and Communicative Action*, Oxford: Blackwell.

Hannerz, U. (1990) 'Cosmopolitans and locals in world culture', in M. Featherstone (ed.) *Global Culture: Nationalism, Globalization and Modernity*, London: Sage.

Harvey, D. (1990) *The Condition of Postmodernity*, Oxford: Blackwell.

Harvey, D. (1996) *Justice, Nature and the Geography of Difference*, Oxford: Blackwell.
Hewitt, K. (1983) 'The idea of calamity in a technocratic age', in K. Hewitt. (ed.) *Interpretations of Calamity*, Risks and Hazards Series 1, London: Allen and Unwin.
Homan, J. (2001) 'A culturally sensitive approach to risk? "Natural" hazard perception in Egypt and the UK', *Australian Journal of Emergency Management* 16 (2), 14–18.
House of Lords Select Committee on Science and Technology (2000) *Science and Society*, Third Report, London: HMSO.
International Federation of Red Cross and Red Crescent Societies (IFRC/RC) (1999) *World Disasters Report*, Geneva: IFRC/RC.
Locke, S. (1999) 'Golem science and the public understanding of science: from deficit to dilemma', *Public Understanding of Science* 8, 75–92.
Macnaghten, P. and Urry, J. (1998) *Contested Natures*, London: Sage.
Maskrey, A. (1989) *Disaster Mitigation: A Community Based Approach*, Development Guidelines No. 3, Oxford: Oxfam.
Maskrey, A. (1995) 'The Alto Mayo Reconstruction Plan, Peru – an NGO approach', in Y. Aysan, A. Clayton, A. Cory, I. Davis and D. Sanderson (eds) *Developing Building for Safety Programmes: Guidelines for Organizing Safe Building Improvement Programmes in Disaster-Prone Areas*, London: Intermediate Technology, London.
Millar, R. and Wynne, B. (1988) 'Public understanding of science: from contents to processes', *The International Journal of Science Education* 10 (4), 388–398.
Mitchell, J.T. (2000) 'The hazards of one's faith: hazard perceptions of South Carolina Christian clergy', *Environmental Hazards* 2, 25–41.
Negus, Y. (1992) 'Science within Islam: learning how to care for our world', in F. Khalid and J. O'Brien (eds) *Islam and Ecology*, London: Cassell.
Oliver-Smith, A. and Hoffman, S. (1999) *The Angry Earth*, London: Routledge.
Pelling, M. (2001) 'Natural disasters?', in N. Castree and B. Braun (eds) *Social Nature: Theory, Practice and Politics*, Oxford: Blackwell.
Petts, J., Homan, J. and Pollard, S. (forthcoming) *Participatory Risk Assessment*, Bristol: Environment Agency.
Phillips, M. and Mighall, T. (2000) *Society and Exploration Through Nature*, Harlow: Pearson.
*Qur'an, The Holy* (1993) Translated with notes by N.J. Dawood, London: Penguin.
Rabinow, P. (1986) 'Representations are social facts: modernity and post-modernity in anthropology', in J. Clifford and G.E. Marcus (eds) *Writing Culture: The Poetics and Politics of Ethnography*, Berkeley: University of California Press.
Randall, C. and Bowcott, O. (1987) '17 killed in hurricance devastation', *Daily Telegraph*, 17 October, p. 1 and p. 32.
Sayer, A. (1992) *Method in Social Science: A Realist Approach*, London: Routledge.
Schmuck, H. (2000) '"An Act of Allah": religious explanations for floods in Bangladesh as survival strategy', *International Journal of Mass Emergencies and Disasters* 18 (1), 85–95.
Simmons, I.G. (1993) *Interpreting Nature: Cultural Constructions of the Environment*, London: Routledge.
*The Times* (1987) 'The timetable of havoc in Britain's worst storm', 17 October, p. 24.
Urry, J. and Macnaghten, P. (1995) 'Towards a sociology of nature', *Sociology* 29 (2), 203–220.

Weber, M. (1992) *The Protestant Ethic and the Spirit of Capitalism*, London: Routledge.

Wynne, B. (1991) 'Knowledge in context', *Science, Technology and Human Values* 16, 111–121.

Zaman, M.Q. (1999) 'Vulnerability, disaster and survival in Bangladesh: three case studies', in A. Oliver-Smith and S. Hoffman (eds) *The Angry Earth*, London: Routledge.

# 10 Understandings of catastrophe

## The landslide at La Josefina, Ecuador

*Arthur Morris*

### Introduction

In Ecuador the cities of the Sierra are located at an altitude of between 2,500 and 3,000 metres, with most being the centres of fertile agricultural basins. They are all exposed to the dangers of landslides and *lahars*, huge flows of mud, ice and snow provoked by sudden melting of snow caps at levels above 4,500 metres. A large landslide in the mountains of southern Ecuador, near the city of Cuenca, is the subject of the present discussion. The principal focus is on reactions to the landslide and the apportionment of blame for it. One key question is whether this disaster and others like it should be treated as 'social nature' – a combination of human and natural forces – or as a natural event which cannot be avoided given the environmental conditions. In terms of policy, this is important. Is it possible to adopt a positive policy of landslide prevention, knowing how mankind relates to nature, or should the focus be on adaptation and reactions to each event? Here, we first describe the landslide event, examine possible causes and report on local actors' apportionment of blame for the slide. Finally, the role of different agencies after the event is reviewed, giving some guidance as to how future events of this kind may be dealt with.

### La Josefina

Cuenca city lies at around 2,500 metres, enclosed by high mountains reaching to over 4,000 metres, one of a necklace of such intermontane basins which reach from Quito southwards to the Peruvian border. The area around the city has a dense rural population, working small farms of only one or two hectares, in subsistence farming based on maize but with a variety of animals and vegetable crops. The basin is drained by the Paute river and its tributaries (Figure 10.1). Leading out eastwards to the Amazon by cutting through a major mountain ridge in a valley of near-gorge-like proportions, this river has a powerful flow of 1,200 cubic metres per second, and carries a high sediment load from its strong natural erosion on the slopes of the upper tributaries.

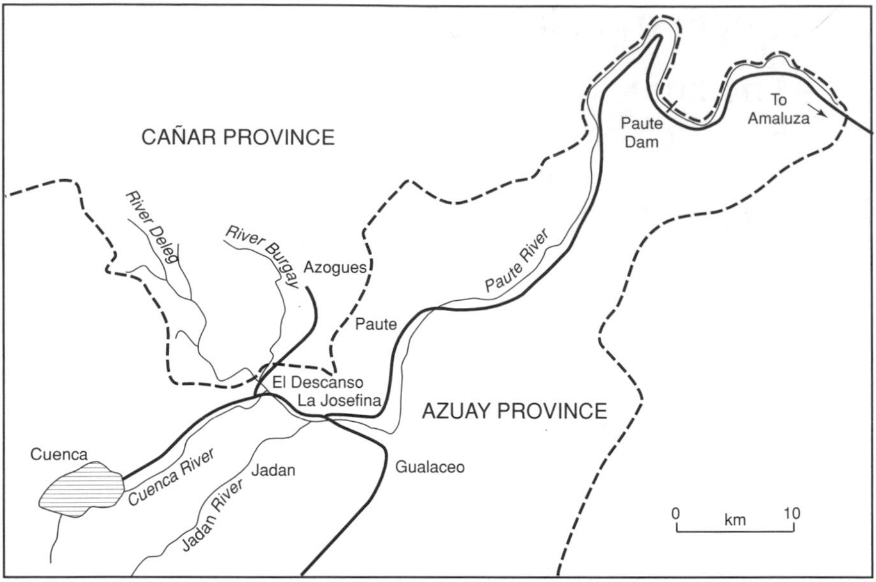


*Figure 10.1* La Josefina.

In 1993 there was a huge landslide at the entrance to the gorge section, which crossed the whole valley and dammed the river, impounding it for 33 days until released. Many people were killed in the initial landslide, perhaps 200. An area reaching around one thousand hectares, including 300 hectares of arable land, was flooded for the period of the damming, causing heavy material damage. The final release of the water caused further damage downstream along the Paute river for 20 km. At one time it was thought to threaten even the Paute hydroelectric dam at Amaluza, some 72 km downstream. The possible threat to the dam brought in interests at the national level, because this dam provided up to 60 per cent of national power generation at the time of the disaster. This was a world-scale event, among the four or five largest landslides in modern times (Jones 1992), and has not been paralleled in Ecuador or neighbouring countries in subsequent years, though the risk of similar events will always be present. It affected a whole slope from 2,270 metres up to 2,900 metres, along a front of up to one kilometre.

Landslide material dammed the two tributaries of the Paute river, the Cuenca and the Jadan. At first two lakes were formed, in the two tributary valleys, but these coalesced into one after 20 days. The public was made highly aware of the environmental disaster, not just because of media coverage but also because the new lake cut off the main road from Cuenca to the capital, Quito, that provided access to the lowlands to both east and west. Overall, some 40 kilometres of surfaced road and 20 kilometres of unmade roads were flooded and unusable (*El Mercurio* 1993). The railway to Quito was flooded over 43 kilometres. This branch line had always been uneconomic, and was kept open as a welfare service to the villages along its line more than as a commercial railway, and it was closed by the flooding never to reopen. A 20 megawatt thermo-electric power station near the dam, at El Descanso, was flooded releasing oil into the lake, adding an industrial pollution hazard to the existing hazard from Cuenca's urban wastes which were released untreated into the water system. Farmlands, six factories, and four agro-industrial establishments were affected by the flooding.

Contaminated water accumulated, well documented by reporters from Cuenca, over a period of several weeks. This led to pressures for action to open the dam, and unsuccessful efforts were made by an army unit aided by local agencies. Decision-making was made difficult because the arguments for rapid action to drain the lake by upstream residents were matched by the fear downstream that sudden release of the dammed waters would itself be a problem, both for farmlands and for the hydroelectric Amaluza dam (Monsalve 1993). The best hope was that the dam would clear naturally, and gently, through the pressure of the water when it reached the crest.

The dam refused to give way naturally, and a discharge channel was excavated by army engineers. When this failed because of continuous falls of loose rock, explosives were used to clear the channel. The maximum flow of the torrent now released was estimated at nearly 10,000 cubic metres per second. This destroyed several bridges and flooded the small town of Paute and neighbouring intensive farming areas. No direct damage was caused to the Amaluza dam. However, there was some damage caused by the chief engineer at the dam! The day after the disastrous release, the national president of Ecuador made a visit to the area and the turbines were opened up at a time when the water was still turbid from the release upstream. This damaged the injectors to the turbines which then had to be closed for repairs. In all, around 200 persons were killed directly by the floods, and 14,000 persons displaced with damage to land and buildings. Flood damages to agricultural lands, factories and residential infrastructure must have added up to many millions of pounds. More intangible are costs that will stem from lost production as agricultural land and the capacity of the Amaluza dam have been compromised over the long term.

## Who to blame? Alternative views

### *The scientific discourse*

The prestige attached to scientific accounts of the disaster, informed by Western science, made them the dominant discourse regarding the event, though alternative ways of looking at the disaster were accepted by different cultural groups as is explained below.

In the scientific account a first direct cause must be the road which ran parallel to the river and which was buried by the slide. This road was cut into the foot of the slope, destabilizing it over a period of time. Commercial quarrying also cut into the slope at this point (Monsalve 1993: 43–44). Immediate direct causes were the heavy rain that had occurred over the previous month, and a possible earth tremor on the night of the landslide. Runoff from the rainfall was thought to have entered into large crevasses formed at the head of the slope, lubricating the mass and allowing it to move partially as a single body down a weak plane (Jaramillo, in Galarza and Galarza 1993). Long-term, indirect causes of the catastrophe are listed as, first, local geologic structure, with less stable and more permeable rocks dipping steeply towards the river, overlying hard, impermeable rocks at the base; second, the erosive strength of the river below; third, steep slopes; and fourth, deforestation (ibid.).

For the apportionment of 'blame' for the event, it is relevant to distinguish natural and human causes. The natural causes make a formidable array, seemingly pointing to a permanent danger of landslides. Qualitative observations by the writer support this view. For some 20 kilometres below the landslide site, down to the settlement of Palmas, there are scars of about thirty-five former slides. Using vegetation re-growth rates as an indicator, six of these slides appear to have been recent, with the majority from the last 100–200 years at the most. Similar slides have occurred since La Josefina. In the year after the Josefina slide, a tributary of the Paute, the Collay, was cut off and dammed, although this dam cleared itself naturally a week later (*El Universo* 12 October 1994). At the same time, mass movements of material on the left bank of the river, at the site of the Amaluza dam, required slope stabilization works to be implemented to protect the dam.

Turning to the human causes of La Josefina, the undercutting of the road at the slide site, together with quarrying, with consequent unloading of the whole slope, were accepted by the scientific community as the immediate causes of the accident. Loss and degradation of the vegetation above and around the landslide site is another human cause, though it is uncertain to what extent forest cover might have prevented the slide.

### *Science and politics: the political ecology view*

The above is a standard view that informed opinion in Cuenca. There is an alternative scientific discourse based on land use in this region. It did not surface at Cuenca, but its regular use elsewhere makes it worth considering here. Poor farming practices leading to soil erosion, excessive pressure on the land resource, failure to maintain the forest cover on steep slopes, may all be indicted as reasons for a high general landslide risk throughout the region. The whole Inter-Andean corridor region of Ecuador has been mapped by ORSTOM as an area where natural soil erosion is high, and as subject to a high risk of intensified erosion through human agency (de Noni and Trujillo 1989). Behind the soil erosion explanation there is the implication of farming malpractices, and the blame for these is given to modern capitalist types of farming, with little concern for the long-term conservation of the environment. In pre-colonial times, the view is that indigenous people managed the environment better, limiting any negative effects of natural events. Vulnerability analysis applied to Peru, but relevant too for Ecuador, shows that pre-Hispanic land use and building technology limited the extent of any disaster (Oliver-Smith 1994). The political ecology argument is broadly that, by contrast, colonial and capitalist exploitation led to current crises of environmental damage, and thus to disaster (Blaikie 1985).

This approach is not fully convincing for Latin America. Over the last 50 years land reforms, anti-capitalist in nature, have sought to break up the old estates. But in countries such as Ecuador these have not reversed poor capitalist farming practices. Instead, around Cuenca, land has been divided into excessively small plots and is worked by farmers from other areas with little experience in actually managing land over the long term. A number of agrarian reform areas around Cuenca, and in the province of Cañar to the north, have been visited by the researcher on different occasions in the 1990s (Morris 1997), and it is evident that the new owner-farmers come with little experience and are over-intensive (too many cattle, too much clear tilled maize, farms too small). In contrast, Preston has shown that in Tarija, Bolivia, the erosion levels achieved by small farmers are not exceptional and not a problem (Preston 1990, 1997, 1998). But Tarija is an unusual area, experiencing heavy outmigration to nearby Argentina, with declining farmer pressure on the land.

If there is an indictment of capital, it is very indirect. In the 1950s, there was a collapse of the main craft industry of Cuenca, Panama hat-making, due to incipient globalization and the identification of alternative hat sources in the Far East. True Panama hats became too expensive for the market of the day. This created an economic crisis which may have led to greater pressure on the land as craft workers were forced back into farming. But even this tenuous line of linkage is doubtful, as the hat-makers were a class moving out of agriculture and into craft work and, eventually, urban jobs. Few are likely to have intensified their use of the land.

### *A third way: indigenous knowledge*

Yet another way of looking, still in scientific mode, at the human–land relationship is through an understanding of local or indigenous knowledge, which has recently come into vogue. Academically the use of this knowledge is attractive too, because it engages with the post-structuralists' concerns to identify and advocate alternative views of nature (Blaikie 2001; Warren *et al.* 1995)

In the context of La Josefina, the argument would be that indigenous or local knowledge would prevent similar accidents through a more conservationist approach. But there is an opposing view. In the altiplano of Bolivia/Peru indigenous land management produced extensive terraces and raised field technology which improved soil conditions and micro-climate (Denevan 2001). Some of these features have been reintroduced recently, but in the Cuenca region there is no evidence of inherited knowledge on the best farming practices for such areas. Mono-cultivation of maize in rows following the slope is typical, and this leads to heavy erosion. Such practices as tree cultivation, terrace formation with shrubs and intercropping with legumes are scarcely known among the small farmers who inherited the agrarian reform lands, but, in any case, the slope land along the rivers might be best employed in conservation plantings and not in agriculture at all (Morris 1985). If there is local indigenous knowledge about this it is ignored.

### *Non-scientific discourse*

There were other views. One perspective held by peasant farmers blamed the landslide on supernatural forces; like other natural catastrophes it was really a punishment by a god or gods, for the misbehaviour of humankind (Codevilla 1993). Such a pre-scientific explanation is tied to the minimal level of formal education available to rural inhabitants. Within the tradition of the Cañari people, the dominant indigenous grouping of the region, there are legends about lakes, which have spiritual significance, so that a sacred meaning could be attached to the newly formed lake (Monsalve 1993: 20). More generally, these people have a set of older beliefs, a kind of animism which relates to what is usually now termed the 'cosmovision' of all Central Andean peoples, in which mankind is seen as one more element of nature so that any human action sees a natural response. Natural features such as mountains and rivers are regarded as gods able to support or exact revenge on humans (Howard-Malverde 1981). A landslide could be considered such an act.

Most people in the Cuenca region do have a Western religious education, which puts a somewhat different view of the event, though a good deal of superstition still exists in Latin American Catholicism. In the dominant Catholic understanding, there was no open view of this as divine retribution, but there was certainly recourse to prayer and to favourite saints who might intervene to help the survivors of the tragedy. This is non-scientific, at least to the extent that it has no belief that science could work either to prevent the event or to help survivors.

Another kind of view is the one most discussed in the books written after the disaster (and cited in this chapter) by Codevilla, Galarza and Galarza, and Monsalve. Rather than focusing on the causes of the disaster, it goes immediately to the laying of blame on outsiders, and calling for a variety of regional, national and international institutions to take action as they were to be held responsible. In most countries of Latin America, a kind of dependency has emerged, in which major outside institutions such as the national government, or the aid agencies, are looked to by individuals for material and financial aid. The weakness of civil society in the continent has been identified by Francis Fukuyama (1995) relating back to the organization of power in colonial times and continued on into the present era by corrupt and dictatorial governments. This weakness is associated with a lack of entrepreneurship, and an exploitative attitude by the individual rather than willingness to combine and co-operate. There is even a name for the attitude, *asistencialismo*, which has no precise translation into English but means relying on assistance, help or aid from elsewhere. Sarah Radcliffe (2001) does advance the idea that civil society is reorganizing itself in the Andes. After the 1970–1990 period of weak civil society, she sees a movement towards new structures, notably the ethnic movement starting with the indigenous peoples of eastern Ecuador. However, while this movement has had wide impact it has not affected Cuenca.

This dependent view of blaming outsiders circulated widely in the urban community of Cuenca itself, literate and able to understand the physical nature of the event. Under this view, the predisposing factors of steep slopes, powerful erosive forces, and geology and climate, which combine to make landslides a likely event, are all taken for granted. What remains is the fact that the road and quarrying were allowed by public institutions, and that action after the event should have been in the hands of these institutions and that they have failed. The agencies most relevant are the regional development agency (CREA), present since the 1950s, the Paute Valley Management Unit (UMACPA), the state electricity company (INECEL), the Inter-American Development Bank, which co-financed the Amaluza hydroelectric dam with the Ecuadorian government, and the provincial and national governments.

Summarizing this section, which has examined various views of natural hazard, what becomes evident is the limited utility of following through the differing lines of argument and calls for action. In contrast to the West, where the contest is generally between two sides, one arguing for protection of nature and the other for exploitation (Smout 2000), we have here overwhelming natural forces and human pressures which cannot readily be assuaged.

## Coping with disaster

The critique made by those seeking to put blame on the outsiders can best be seen in relation to the exercise made in coping with the aftermath of La Josefina. The first matter of importance was how to handle the landslide itself. This was always a no-win situation. If the slide was broken up, there was damage downstream; if it was not, there was flooding upstream, and possibly a still greater flood on the eventual release downstream. The intervention of the army came when it became apparent that local authorities were unable to come to a collective decision.

Regarding the time after the event, the three books on the disaster which have been cited, Codevilla, Galarza and Galarza, and Monsalve, all assemble and abstract from a variety of local views, and all identify a lack of co-ordination between agencies such as CREA, the local police and fire services, the Church, the army, and the various local NGOs, some of which had no close link with the landslide but sought to help. The criticisms range from the lack of effective aid from the NGOs, although there were 47 of them in the region, to the lack of objective focus by the provincial politicians, who saw in the landslide a chance for bringing in disaster funds from Quito or abroad, funds which would swell their own coffers and importance. A Crisis Committee was set up, but with 15 members it was too large to make rapid and effective decisions. No real aid reached the affected families – apart from immediate shelter or medical treatment in the first weeks after the disaster – and there were no autonomous mechanisms for providing ongoing welfare services or for restimulating the economy and society of the affected region.

Planes brought 630 tents as a donation from the USA government, for the homeless, but the slow resolution of the problem is shown by the fact that these were still in use two months later. Reconstruction at the town of Paute was eventually allowed in the same flood-prone sites as had previously been occupied. The Inter-American Development Bank did provide US$200,000 in emergency assistance to Ecuador to help the people displaced by the slide. This was used to purchase medicines, bedding, housewares and other materials needed by the 14,000 people evicted from areas downstream, but when this ended there was no ongoing local organization (Inter-American Development Bank 1993).

The official reaction to the disaster was thus a caricature of what is expected internationally and identified by the media as a norm for disaster relief. As put by Maskrey (1994), the national effort is in the 'kitsch paradigm', bringing in food and clothing when these are not really needed, but failing to bring in institutions which would help the rural people over the months and years to come. Codevilla (1993: 50) points out the other side of the organizational problem: no effort was made by local groups of their own accord. In the great majority of cases there was an outbreak of *asistencialismo*, locals simply waiting to be helped. The point I am leading to is that those who criticized the lack of help from the outside are to a great extent criticizing themselves. There was no grassroots movement, no structure in place which could handle this disaster with confidence. If the dominant discourse was that of science and intervention by mass force (the army intervening to break the landslide dam and doing little else), it was because of a lack of alternatives.

## A practical view: identifying the problem, planning for the future

We may isolate from the above review of possible culprits and faults, some real problem areas, and also detect a failure to identify causal forces. In this section an attempt will be made to look at positive planning for the future, based on the failures of the recent past.

While mention was made by many commentators of the huge natural forces, there is no extension from this kind of comment to its logical conclusions. Naturally high levels of erosion and sedimentation are a feature of these high mountain environments. In this situation, events like La Josefina must be expected at regular intervals and should not be regarded as outlandish. As was noted above, the valley below this site is scarred for miles with similar landslides from the past. This understanding of a natural event will release local society from the need to cast major blame on any human intervention. Instead of seeking to avoid the major event, which is unpredictable in space and time, the focus of effort might well be placed on organizing the management of disaster relief and regeneration of affected areas through a new organization whose specific role that could be.

There could also be a move away from a hard engineering approach (trying to match the power of nature) to soft engineering (looking at the environment and making modest, low-energy adaptations of our use of it). Soft engineering has been recommended in a variety of situations, including Britain during the floods of the year 2000. At a world level, Abramovitz (2001) advocates maintenance of healthy and resilient ecosystems as the key, replacing structural engineering which often fails in disastrous manner. In the context of Ecuador and the landslide hazard, the soft engineering approach involves recognition of the huge natural forces, which cannot be totally controlled, in the Andean environment. What can be done is to manage the physical environment through adaptation of the major land uses, notably agriculture. Very little mention of this was made after La Josefina, but in fact the land pressure problem is intense, as noted earlier. Despite there having been two attempts at land reform in the country, in 1964 and 1973 (Barsky 1988), and a developmentalist reform in 1994, the regional problem of landholding in the Cuenca region remains critical. One arm of the soft engineering approach is greater planning control over land use (Abramowitz 2001), limiting intensive farming and encouraging forestry in critical zones. This was a conclusion also drawn specifically for highland Ecuador by the present writer (Morris 1985, 1997). However, such proposals depend on the real ability to manage and control land use in the rural areas. This has never been really possible in recent time, and would require a more extensive set of powers given to planners, who must face a large farmer population dependent on being able to work their tiny farms at maximum intensity.

Farm size and farm operation are topics of critical importance in our view; over 50 per cent of the farms in the Paute basin are less than one hectare, or 2.5 acres, in size. These are *minifundios*, farms which must be worked intensively if they are to provide their owners with enough to eat. A production problem is that the typical crop, grown on all subsistence farms, is maize, normally grown as a clear-tilled crop where no weeds are allowed. On test erosion plots used by UMACPA, the soil erosion on steep slopes with maize as monoculture reaches over 200 tonnes per hectare per annum.

This is greatly reduced, by more than half, when other crops, typically beans or squash, are intercropped with the maize; it is still further reduced by the use of simple anti-erosion practices, such as terraces and cultivation across the slopes instead of the standard of the region which is up and down the slopes. Unfortunately, only a small minority of farmers actually engage in such practices.

Agrarian reform was supposed to have brought relief to the system by making more farmland available to those who worked it; instead, the land remains critically little, as there are always too many farmers and potential farmers for the good land available, as is still the situation throughout the Central Andes. Estates in the Cuenca region (Barsky 1988: 76) were divided into units under four hectares, bigger than the regional average but still insufficient for commercial farming. Farm population is unknown, but the rural population has not declined over the past two decades. From observations by the researcher, much marginal land for cultivation, in the zone around 3,000 metres used for cattle pasture before the reforms, has been put to cultivation in the recent past, adding an erosion source to the existing ones. There is further evidence of the intensification of farming, in the comparison of land use maps, made using the 1:100,000 scale maps of two time points – 1977 (CREA 1980) and 1993 (UMACPA 1993) – which indicate a rapid conversion of woodland to pasture and of pasture to cultivated land.

Additionally, the agrarian reform laws resulted usually in the distribution of estate land to new farmers, but little or nothing in the form of new infrastructure such as farm to market roads, or training for those who now had to manage their own land. Farmers have not generally learned conservation methods.

The more recent, neo-liberal Agrarian Development Law of 1994 or Ley de Desarrollo Agrario, covered in Parra (1994), took the place of the agrarian reform, revoking the earlier legislation but doing little to change the situation. For large farmers, mostly to be found in other regions of Ecuador, there was a new protection in the declaration of the rights of private owners to keep their land, in contrast to the previous permanent threat of expropriation and distribution to peasants or estate workers. But in this region, where estates had mostly been previously split up into co-operatives, the preferred form under the 1970s agrarian reform, most farming was already done by individuals, not by the co-operative which had been created simply to acquire the land. The neo-liberal reform of 1994 only confirmed this situation, without changing the farmers' economic situation or pressures. One of its dispositions, Article 34 (Parra 1994: 24), did state that *minifundios* should be combined together or formed into co-operatives, as in the 1970s, to achieve reasonable size of farm units, but no forceful action was taken to achieve this.

It is thus the case that the farming sector, the main user of land in rural areas, is dysfunctional for conservation and for the limitation of erosion. The most relevant agency which might be expected to have changed the scene is UMACPA. It was set up originally because of the fears of INECEL and the IDB that excessive erosion would threaten the existing Amaluza dam's period of useful hydroelectric production, and would compromise a further dam which was being planned. Conservation was at the core of their programme, which includes demonstration farms and plots to show conservation methods, crop rotations, crop and animal combinations, the use of trees and shrubs, and similar techniques. UMACPA, like CREA, has encouraged such programmes as the setting up of tree nurseries attached to primary schools, where pupils might learn the value of trees and the need for their care. It has also encouraged small nurseries generally in the region, to stimulate more planting for conservation and production purposes. These nurseries have been received with limited enthusiasm. Those visited by the writer in 1992 and 1994 were too small to have any real effect on afforestation projects, and no capital was put into their establishment. On the small farms there is little extra land for planting trees, and trees take land away from livestock or crops, as reported for similar programmes in Ethiopia (Fitzgerald 1994). Planting of a native alder, *Alnus Jorullensis*, which maintains nitrogen in the soil, does give the possibility of combining pasture production with timber production, but the plots used for this crop are tiny and experimental.

A large recent programme of UMACPA has been to designate Protection Forests, *Bosques Protectores*, which supposedly restrict land use in the headwaters of the river basin and maintain native forests. This programme defined 12 large forests having areas of 15–30 km diameter each under the control of a single ranger. Such a programme, if carried through, would have benefits for the long-term planning of sectors such as forestry and forest products (Morris 1985, 1997), but it has been aspirational rather than practical and there is no way in which vigilance over land use can be exercised.

There is another relevant regional agency, CREA (Comisión para la Reconversión Económica del Austro), the Redevelopment Commission for the Southern Region, in existence since the mid-1950s. It also has conservation programmes, and programmes for the planting of timber trees and agroforestry, alongside its general programme for diversification of the economy and the development of new industry. In one or two areas, notably in canton Giron south of Cuenca, nursery development has gone hand in hand with fairly intensive afforestation that has reached commercial scales, and is now self-supporting as the forest owners provide a constant demand for young trees. On the other hand, the fact that the farming problem has intensified during its regime is one indicator that it has not had great success in the country areas.

## Conclusions

Our findings with regard to the way such an accident may be viewed are somewhat pessimistic. Although a variety of *Weltanschauungen* do inform different views of the disaster in Cuenca, different scientific knowledges from local or indigenous knowledge have not been useful here, any more than have pre-scientific discourses from the region based on the ancient cosmovision. The capitalist critique attached to the situation of farming scarcely applies in a region where reform has been passed through but land pressures remain. At another level, the views centred on critique of local organizations and their response are a critique of civil society in the region, not of the lack of specific action. This leaves us with little more than the need to adjust to natural forces, and to have recourse to scientific information and modern technology for countering this kind of disaster.

The final assessment regarding La Josefina is that there is no solution which can totally prevent individual catastrophic events. There are, however, actions which could and should be taken for the mitigation of disaster effects. Most responsibility lies with the national government, which could change the context by encouraging rural development and rewarding farmers with better prices and help for conservation, leaving local agencies to concern themselves with disaster relief.

## References

Abramovitz, J.N. (2001) 'Averting unnatural disasters', in L.R. Brown, *State of the World 2001*, London: Earthscan.

Barsky, O. (1988) *La Reforma Agraria Ecuatoriana* (2nd edition), Quito: Editora Nacional.

Blaikie, P. (1985) *The Political Economy of Soil Erosion in Developing Countries*, Harlow: Longmans.

Blaikie, P. (2001) 'Social nature and environmental policy in the South: views from verandah and veld', in N. Castree and B. Braun, *Social Nature: Theory, Practice and Politics*, Oxford: Blackwell.

Codevilla, U.R. (1993) *Antes que las aguas nos alcancen*, Cuenca: Rumbos.

Comisión para la Reconversión Económica del Austro (CREA) (1980) *Utilización del suelo y de los paisajes vegetales de la Sierra*, Quito: CREA.

Denevan, W.M. (2001) *Cultivated Landscapes of Native Amazonia and the Andes*, Oxford: Oxford University Press.

*El Mercurio* (1993) [Cuenca daily newspaper], 19 April.

*El Universo* (1994) [Ecuador national newspaper, Quito] 12 October.

Fitzgerald, M. (1994) 'Environmental education in Ethiopia', in A. Varley (ed.) *Disasters and Development*, Chichester: Wiley.

Fukuyama, F. (1995) *Trust: The Social Virtues and the Creation of Prosperity*, London: Hamish Hamilton.

Galarza, J.Z. and Galarza, L. (1993) *Mas allá de las lágrimas*, Cuenca: Ninacuru Editores.

Howard-Malverde, R. (1981) 'Dioses y diablos; tradición oral de Cañar, Ecuador', *Amerindia, revue d'ethnolinguistique amerindienne*, Paris.

Inter-American Development Bank (IDB) (1993) 'Urgent aid follows landslide', *The IDB* (June), 11.

Jones, D.K.C. (1992) 'Landslide hazard assessment in the context of development', in G.J.H. McCall, D.J.C. Laming, and S.C. Scott, *Geohazards: Natural and Manmade*, London: Chapman and Hall.

Monsalve, L.R. (1993) *La tragedia del Austro*, Cuenca: La Golondrina.

Maskrey, A. (1994) 'Disaster mitigation as a crisis of paradigms: reconstruction after the Alto Mayo earthquake, Peru', in A. Varley (ed.) *Disasters and Development*, Chichester: Wiley.

Morris, A.S. (1985) 'Forestry and land-use conflicts in Cuenca, Ecuador', *Mountain Research and Development* 5 (2), 183–196.

Morris, A.S. (1997) 'Afforestation projects in highland Ecuador; patterns of success and failure', *Mountain Research and Development* 17 (1), 31–42.

de Noni, G. and Trujillo, G. (1989) 'Quelques réflexions au sujet de l'érosion et de la conservation des sols en Equateur', in *Colloques et Séminaires; Equateur*, Editions de l'ORSTOM, Institut Français de Recherche Scientifique pour le développement en cooperation, Paris.

Oliver-Smith, A. (1994) 'Peru's 500 year earthquake: vulnerability in historical context', in A. Varley (ed.) *Disasters and Development*, Chichester: Wiley.

Parra, W.O. (1994) *Normatividad Jurídica Agraria*, Cuenca: Ediciones Sela.

Preston, D. (1990) 'From hacienda to family farm: changes in environment and society in Pimampiro, Ecuador', *Geographical Journal* 156 (1), 31–38.

Preston, D. (1997) 'Reevaluating the sustainability of contemporary farming systems', in T. van Naerssen, M. Rutten and A. Zoomers (eds) *The Diversity of Development: Essays in Honour of Jan Kleinpenning*, Assen: Van Gorcum.

Preston, D. (1998) 'Post-peasant capitalist graziers – the 21st century in southern Bolivia', *Mountain Research and Development* 18 (2), 151–158.

Radcliffe, S. (2001) 'Latin America transformed: globalization and modernity', in N. Castree and B. Braun (eds) *Social Nature: Theory, Practice and Politics*, Oxford: Blackwell.

Smout, T.C. (2000) *Nature Contested: Environmental History in Scotland and Northern England Since 1600*, Edinburgh: Edinburgh University Press.

UMACPA (Paute Valley Management Unit) (1993) Convenio INECEL-CLIRSEN, 'Cobertura y uso de la tierra', Unpublished document, Cuenca, Ecuador.

Warren, D.M., Slikkerveet, L.J. and Brokensha, D. (1995) *The Cultural Dimensions of Development*, London: Intermediate Technology Publications.

# 11 Vulnerability reduction and the community-based approach

## A Philippines study

*Katrina Allen*

## Introduction

Globally, there are many regions that, like the Philippines, are subject to frequent extreme events such as typhoons. The occurrence of these events is an integral part of the seasonal cycle for affected communities and their members. In this respect, those affected manage levels of vulnerability according to their priorities and capacities as part of their daily existence. Vulnerability to 'disasters' can only be fully understood and addressed through the consideration of everyday livelihoods and underlying vulnerability (described below). Vulnerability is too closely tied to societal and environmental processes of development and change to be treated as a separate phenomenon in times of crisis (Hewitt 1983; Winchester 1992). Among the contributors to *local* vulnerability are factors and processes that have far wider resonance and origins. These include market forces and policy trends. Addressing such issues within the context of micro-level projects poses a significant challenge for proponents of community-based approaches.

Before proceeding further I shall clarify my use of the terms: 'vulnerability', 'underlying vulnerability' and 'community'. *Vulnerability* is defined as a degree of susceptibility to the effects of events or shocks, of processes of change or of a combination of factors, including stresses, which is not sufficiently counterbalanced by capacities to resist negative impacts in the medium to long term, and to maintain levels of overall well-being. Vulnerability is manifested as a limited or lessened ability to cope with potential or actual situations that may arise. In contrast, *underlying vulnerability* is experienced as a contextual weakness or susceptibility underpinning daily life. The term *community* is employed to describe residents of the geographically and politically defined unit of the *barangay*.[1] As is true for any of the social units of which societies are comprised, local communities in the Philippines are subject to high degrees of diversity, both internally and between different community units. This is not to imply that communities are always too diverse to be capable of unified decision-making or action, simply that – like any social unit – communities have their limitations and weaknesses (Midgely 1986; Wade 1986).

The community-based project at the centre of this study is part of an initiative of the Philippine National Red Cross (PNRC) called the Integrated Community Disaster Planning Programme (ICDPP).[2] This programme is intended to reduce vulnerability to disaster events using a community-based approach. It represents an important (if partial) shift in emphasis within PNRC, which has parallels within the global disaster management field. This trend has two elements. First, emphasis is moving from disaster response to preparedness and mitigation; second, from service provision to the facilitation of community-driven initiatives.

Community-based approaches are held by their advocates to be suitable mechanisms for grasping the dynamics and complexity of vulnerability, as manifested at the local level, for addressing vulnerability and strengthening local capacities. This chapter explores the key factors constraining adherence to this ideal in practice, with reference to vulnerability reduction initiatives implemented under the ICDPP project in *barangay* Tigbao. I contend that the primary barriers to effective vulnerability reduction have their roots in the conceptualization of vulnerability within a disaster management context, and in the manner in which constructions of vulnerability are translated into project initiatives.

I begin by outlining the theoretical and pragmatic reasons why community-based approaches are increasingly upheld as effective and appropriate mechanisms for reducing vulnerability. I then present the key limitations of such approaches in practice. In particular, I focus upon differential conceptualizations of vulnerability among project actors, and upon the role this plays in limiting the scope and effectiveness of vulnerability reduction strategies.

## *Barangay* Tigbao

The Philippine archipelago lies in the circum-Pacific typhoon belt. The country experiences an average of 20 typhoons a year, eight to ten of which are highly destructive (PNRC 1994). Tigbao is situated on the island of Leyte in the Eastern Visayas region of the Philippines. Tigbao is a *barangay* in the municipality of Libagon, Southern Leyte province. Over the last decade, Southern Leyte has experienced destructive typhoons on an almost annual basis (Southern Leyte Province 1997). Tigbao is a coastal settlement, prone to destructive storm surge during typhoons. Flash flooding of hillside streams causes damage to infrastructure and to agricultural lands. Although historically there have been few deaths, and typhoon experiences are mild by Philippine disaster standards, the livelihood impacts are significant.

### *Life and livelihoods in Tigbao*

The main settlement areas of Tigbao occupy a narrow, low-lying strip between the shoreline and steep hillsides. The combination of population pressure, limited availability of land for house-building and cultivation, migration from upland areas and coastal erosion has contributed to the encroachment of housing upon the exposed foreshore area. Most of the available low-lying land is used for growing rice, while the hillsides are cultivated with coconut palms, abaca (hemp), bananas and rootcrops.

Migration to urban centres in search of employment is common, particularly among *barangay* youth. Availability of livelihood options for large sectors of the population is low. One prominent resource-poor group with a high level of social marginalization is that comprising those who have migrated within the last generation or two from upland areas. These migrants reside in the coastal/foreshore area of Tigbao, developed largely to accommodate their influx during times of insurrectionist–military clashes in the upland areas. Rice farming is small scale and is, for the majority, more for own consumption rather than for significant commercial production. Tigbao has both landowning and tenant farmers, with a large proportion of landless labourers who tend to undertake manual work in agriculture as well as in the construction industry. Many community members are involved in fishing. There exists a small minority of resident salaried professionals, such as teachers or local government staff. More work as drivers, carpenters, artisans, traders and storekeepers. Most households rely upon a variety of livelihood sources. Livelihood diversification is a form of coping strategy, or mechanism to alleviate vulnerability (Chambers 1993), which relies principally upon the capacities and access to resources (such as land, labour, education, credit and cash income) of the individuals, households, kinship networks and cooperative groups of which the community is comprised. A summary of community-level vulnerability experiences is presented in Figure 11.1. Note that these encompass the (interlinking) social, environmental, political and economic spheres of community life.

### *Local leadership*

Each *barangay* in the Philippines has an elected captain. Considerable decision-making power and status is attached to the position, and prominent local families often vie to place one of their members in this role. All the significant problems and disputes within the *barangay* are brought to the captain, who also negotiates and co-ordinates with government officials, fellow *barangay* captains of the municipality and NGO staff. Captains are expected to campaign to bring funding from outside agencies into community-level projects. They are strongly judged on the basis of their achievements in this respect. Together with their appointed councillors, they form the various sectoral committees required by local government regulations – for example, for disaster co-ordination. *Barangay* officials are responsible for the budgeting and allocation of scarce *barangay* resources and for the dissemination of information to community members. As such, they are key players in community-based initiatives, often acting as representatives of the *barangay* community.

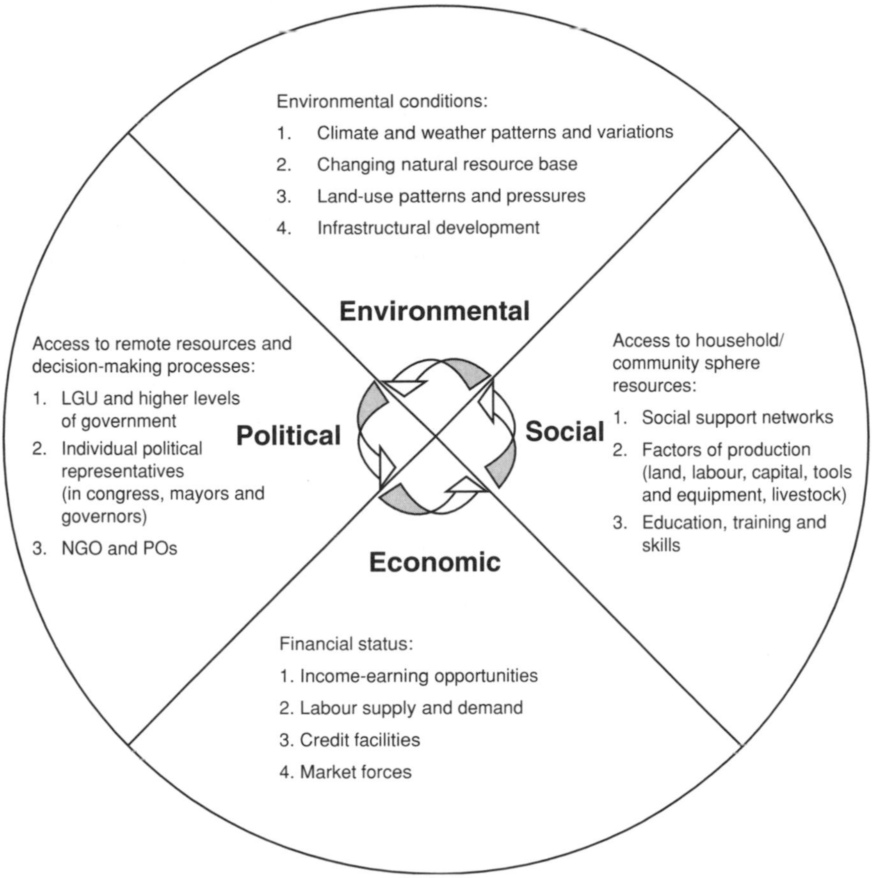


*Figure 11.1* Overview of community-level vulnerability experiences.

### *The Red Cross project*

ICDPP is an experimental programme of PNRC, funded by Danish Red Cross. The ICDPP project in Tigbao was initiated in 1997, with the main activities being carried out from 1998–1999. The project in Tigbao incorporates elements of information dissemination, as well as training in first aid and disaster management strategies, with a focus upon preparing for disaster events. Training culminated in the formation of a voluntary *Barangay* Disaster Action Team (BDAT). The project has implemented the following mitigation measures:

1 building a seawall;
2 strengthening an existing river dike, dredging and diverting the course of a stream away from agricultural land;
3 planting mangroves to protect the eroding coastline and provide habitat for marine life.

ICDPP requires substantial co-ordination with and between members of local government at provincial, municipal and *barangay* levels. Long before the Tigbao project had been envisaged, ICDPP staff were involved in consultation and negotiation with government officials as to the purpose of the programme, the nature of proposed projects, and as to which local government resources could be drawn upon and to what extent. The project site itself was selected by PNRC (on the basis of its vulnerability rating according to ICDPP-defined criteria) from among the 14 *barangays* of the municipality. Under ICDPP, the involvement not only of community members but also local government bodies is considered fundamental in promoting the long-term sustainability, replicability and effectiveness of such projects. At the community-level, the *barangay* captain has played an active role in shaping project outputs, organizing community members, disseminating information, sustaining activities, reporting progress and co-ordinating with ICDPP and local government officials. The inputs of community members (including trained BDAT members) are subject to existing hierarchical arrangements, with the authority of the captain – and ultimately the mayor – required for them to proceed legitimately.

In the project, aspects of social vulnerability are addressed only within a strictly limited framework, while the project focuses upon preparing for physical events. This approach contrasts with community member representations of typhoon-related hazards as integrated with everyday life and livelihoods, as illustrated in Figure 11.1.[3]

## Towards a community-based approach

The use of community-based approaches to alleviate vulnerability in a disaster management context is increasingly endorsed (Hall 1997; Lewis 1999; Luna 2000; Twigg 1998). Community-based approaches are considered both appropriate and effective mechanisms for targeting local vulnerability – not least because it is at the local level that the effects of natural hazard events are experienced and that vulnerability is manifest. There are compelling pragmatic grounds for addressing vulnerability at this level, and, as explored below, these have theoretical underpinnings.

### *Theoretical underpinnings*

Theorists such as Blaikie *et al.* (1994), Burton *et al.* (1978), Hewitt (1983, 1997), Watts (1983) and White (1974) – and the majority of contributors to this volume – assert that disasters are essentially social happenings, created by the interplay of natural and social factors and processes. In order to address vulnerability to such events, an understanding of their social context is required (Hewitt 1983; Twigg 1998; Winchester 1992). This social context incorporates not only those factors directly impacting at the local level, or under local-level control, but also includes wider factors and processes of development in so far as these influence the actions, capacities or priorities of local actors. An example drawn from my case study illustrates this point. In Tigbao, many of those households who have settled on or close to the foreshore area, in positions which they understand to be vulnerable to typhoon storm surge and coastal erosion, are first- or second-generation upland migrants. Land allocation and fishing opportunities are largely responsible for migrant settlement in this location. Although the migrants are relatively resource-poor and socially marginalized in the context of Tigbao, most consider themselves overall less vulnerable,[4] with greater opportunities in Tigbao than in their former communities. If we accept that vulnerability to natural hazards and processes of development are inextricably linked, it follows that the areas of disaster management and development should also be related in practice (Blaikie *et al.* 1994; Handmer and Dovers 1996; Lewis 1999).

The above example illustrates another important theoretical juncture – namely, recognition of the role of human agency. A purely deterministic approach assumes that the interplay of natural and social processes alone determines the vulnerability of local people and institutions. In the hazards field, Blaikie *et al.* (1994), Hewitt (1983) and White (1974) reject this type of determinism. In order to understand vulnerability and its causes people need to be seen not as victims but as actors with the capacity to contribute to the management of their vulnerability levels, according to their resource bases, opportunities and choices (Cannon 2000). Human agency not only plays a benevolent role in alleviating vulnerability but also a causal role in determining who is vulnerable, in which aspects of their lives, at which times, and to what extent (Hewitt 1983; Blaikie *et al.* 1994).

### *Pragmatic and ethical underpinnings*

Participatory, community-based approaches have been developed in response to the perceived failure of 'top-down' approaches to deliver project objectives such as the reduction of vulnerability (Chambers 1983, 1993; Korten 1980). Community-based approaches are intended to elicit greater involvement of local actors in project and programme processes. This is in part a consequence of the recognition of local people as capable actors. Community-based approaches emphasize the capacities of local actors, and the value of their contributions. They often promote the 'right' of local people to participate in vulnerability alleviation initiatives (Johnston and Clark 1982; Korten 1980; Uphoff 1991, 1992). In purely pragmatic terms, community involvement in local projects is intended to increase their acceptability and success rate (Chambers 1983; Pretty and Scoones 1995; Shepherd 1998).

Community-based approaches are intended to incorporate and to build upon local knowledge, experience and capacities in a participant-friendly manner. In practice this has often been translated into working with and through existing social institutions (Cernea 1991; Johnston and Clark 1982).

Participation has proved an inherently difficult notion to put into practice, with a dichotomy of methods, and huge variations in terms of extent and levels of inclusiveness (Midgely 1986). There is, however, growing consensus that local participants *should* be included in project processes.

Community-based approaches are intended to be responsive to change. Project structures aim to be flexible enough to incorporate learning experiences as projects proceed. Emphasis is shifted towards achieving success according to community actor's criteria, and away from meeting organizational targets and enhancing prestige (Chambers 1993; Chambers *et al.* 1989; Korten 1980). All of the points above are intended to increase the effectiveness of projects and to increase the likelihood of long-term sustainability, precisely because they are concerned with creating (largely) community-driven projects, sensitive to local-specific factors. In this manner, power and control of project processes pass in some measure to community-level actors.

## Practical limitations of the community-based approach

In practice the community-based approach has been implemented with varying degrees of success. Drawing upon the case study example, I have outlined the main limitations of the community-based approach in the context of an initiative intended to reduce local vulnerability to typhoon-related natural hazards.

### *Red Cross internal factors*

My first set of limitations are concerned with factors internal to the Red Cross society concerned, and with its donor from within the Red Cross movement. The roots of these constraining factors lie in the institutional framework and directives, as well as project and programme processes. The first of these factors is concerned with the evolving role of PNRC, and associated debate as to what the most appropriate role for PNRC might be in vulnerability reduction initiatives. This debate can be framed in terms of a series of options for focus, which – although not mutually exclusive – can be simplified as follows:

1. working in co-ordination with other agents (especially local government bodies); PNRC specializing in its own established areas of expertise such as disaster preparedness and disaster management, with a particular focus upon informal lectures about natural disasters and how they should best be prepared for, and reacted to;
2. expanding into relatively new areas of activity – for example, livelihood programmes and/or measures, to allow for a more encompassing approach to vulnerability reduction; and
3. playing an advocacy role, facilitating the campaigning of community priorities to appropriate agencies.

In as far as PNRC follows the first option, it can be seen as moving into gaps left by government agencies in PNRC areas of specialization. This is more or less the approach presently adopted by PNRC, and is strongly service-delivery oriented. The second option offers a means of addressing vulnerability at the community level, according to desires or needs for wider vulnerability reducing measures, which have been expressed within the community. However, ICDPP was envisaged 'to pilot test a community-based approach to strengthen the capacity of communities, local government and the Red Cross . . . within the disaster management continuum' (PNRC 1999). As such, in its present form ICDPP's agenda is limited from its outset to addressing vulnerability through a focus upon disaster management. The final option is to an extent already being undertaken, although advocacy is not the main focus of current activities. In this respect, Red Cross principles of 'political neutrality' come into play, particularly as the most likely scenario in this advocacy role would involve facilitating *barangay* officials to access higher levels of local government funding. Any advocacy role could also be expected to diminish as local-level capacities for independent action are enhanced.

Another factor limiting the extent to which the project can be community-driven is the standardization of processes and procedures. There are pressures from within PNRC for centralized designs of programme or service elements. This is especially true of the training format (PNRC 1999). This is in part due to staff looking for clarification of their respective roles, but can also be seen as symptomatic of an organization that has remained largely service-delivery oriented. There are strong arguments against using a highly standardized or 'blueprint' approach.[5] One of the most poignant is that standardization tends to enforce the separation of project planning, implementation and evaluatory components. This in turn inhibits staff on the ground and local participants from applying their learning curve throughout the course of the project (Korten 1980). Instead, evaluation tends to be left to the end of a given project and findings are produced to be applied (it is hoped) to future phases of ongoing programmes, or to entirely new projects. Standardization of procedures also removes much of the onus of responsibility from local-level project actors to tailor projects to meet local-specific needs. Community-based projects require large doses of incrementalism. Implementers and participants alike learn as they proceed about what works and what does not (Johnston and Clark 1982; Korten 1980). Putting this learning into practice requires that project management systems retain the ability to adapt to lessons learnt within a relatively short time-frame (Leonard and Marshall 1982).

It is easy for those caught up in the day-to-day running of projects to become focused more upon their physical outputs than upon meeting underlying objectives. This is as true for project processes and management issues, as for capacity-building activities such as training. Within the Tigbao project much effort went into making protective measures such as the sea wall construction participatory, both in the planning and implementation phases. These 'concrete' measures were widely upheld by those involved as *the* success of the project. Training sessions were also highly rated by project staff. Their success was measured primarily in terms of the number of participants. The nature and extent of such concrete outputs, however, have been shaped by programme-related constraints as well as by limitations in the counterparts available from local government sources. There is a tendency to assume that such outputs will substantially benefit the communities involved without critically assessing the actual impacts of any measures taken. Such assessment should be in terms of actual and potential reduction of vulnerability within the community, and also in terms of the appropriateness of any such measures. The tendency to focus upon easily quantifiable project outputs runs contrary to effective implementation of a process-based learning approach (Korten 1980).

One of the problems with the project cycle approach of planning, implementation, monitoring *then* evaluation is that it is associated with a tendency to focus upon organizational learning, to the detriment of participant or project beneficiary learning. Feedback to community members/volunteers at the local level is generally weak. Evaluation has tended to come too late to influence the project itself, although any lessons learnt will be potentially applicable either in future projects or in extensions of the currently running programme. This can be seen as a failing in terms of the purpose of community participation.

The Tigbao project has relied upon a combination of surveys and training. Surveys were used in the preliminary stages of project site selection and planning as a primarily extractive tool. Training, on the other hand, is first and foremost a means of disseminating information to selected community members. The type of training conducted is relatively informal in nature, and is intended to encourage discussion and sharing of ideas as much as possible. Outside the training sessions, however, there has been little scope for sharing and learning that can be practically applied. The structure of the project itself tends to segregate information extraction and information dissemination elements. Reporting is a relatively formal process that can be sporadic. Generally, unless an issue demands urgent action, reports are filed in PNRC offices to be drawn upon during formal evaluatory phases.

Barriers to practical support of these types of initiative are to be found not in the will to support them but in the formal structuring of project processes and procedures and in the manner in which projects are incorporated into services and programmes of PNRC. Donor stipulations have also placed constraints in terms not just of funding particular project elements but also in terms of timing, with the donor requiring punctual reporting of results within an imposed time-frame. This has led to considerable pressure being placed upon the programme and its staff to demonstrate that implementation objectives have been met within their deadlines.

The development of the community-based programme is symptomatic of a hybrid approach currently evolving within PNRC. This approach uses as its founding basis Red Cross history of information dissemination and training, with some principles of community development introduced. In many respects, PNRC remains primarily a service provider, a position which does not sit comfortably with more participatory, community-driven initiatives. I contend that further adaptation of existing PNRC institutions will be required in order to maximize organizational support of community-based initiatives.

### *Factors external to the Red Cross*

Factors external to the Red Cross are concerned with local government and community actors, their interaction, and their expectations of the project. Inevitably different project actors have their own agenda, and these do not always fit easily with one another. The community-based approach may encourage negotiated decision-making; however, this does not negate the role of hierarchical power relations in shaping and weighting eventual decisions and compromises. This is true at all levels, from decision-making within the community to Red Cross and government inputs.

Project actors also have past experience and knowledge of the operations of one another. On this basis, expectations are formed on all sides even prior to project initiation. For instance, to local government officials PNRC acts as an auxiliary service operating with local government consent in specialized areas such as first aid training. Community members know PNRC as a distributor of relief goods. To many, PNRC is first and foremost a charitable organization spending funds to aid needy people. Although PNRC continues to maintain its role in rescue, relief and rehabilitation, community-based disaster management has a different focus. Whereas PNRC's relief and rehabilitation roles are interventionist and aid-oriented, community-based disaster management focuses upon enhancing the capacity of local communities themselves to prepare for and to manage hazards and disaster situations. In Tigbao, labour was carried out by community members themselves on the insistence of project staff, while funding for implementing mitigation measures was solicited from local government. Despite expectations to the contrary, PNRC became significantly less of a 'giver' and more of a 'facilitator' *vis-à-vis* local actors. The political reality remains that such projects carry potential political leverage in terms of credit attributed for successful implementation and for project outputs. This is an inherently difficult situation for any non-governmental organization and is particularly so for Red Cross national societies, which embrace political neutrality as one of their basic tenets.

### *Depoliticization and constructions of vulnerability*

Paradoxically, community-based approaches intended to empower participants can also serve to *depoliticize*[6] issues surrounding vulnerability. This is partly due to project discourse which associates vulnerability with hazard events and treats non-event-centred causes and manifestations of vulnerability as outside the scope of the project. Organizational and government discourses tend to be shaped by their own agendas. For instance, in the Tigbao project vulnerability reduction is treated as a benevolent humanitarian issue, divorced from its social, economic and political context. This corresponds with both PNRC and government approaches to disaster management, and steers the project clear of politically sensitive issues. This means that powerful companies are more able to evade the regulation of activities such as quarrying and logging, which can have severe implications for downstream or floodplain-based communities. Addressing such issues single-handedly is not a role that PNRC could – or should – adopt. This does not alter the need for local vulnerability issues to be addressed within this wider, often politically contentious, context.

The depoliticization of vulnerability issues described above is reinforced in community-based initiatives which allocate the primary responsibility for local vulnerability reduction to community members and officials who lack the jurisdiction or political power to address wider factors and processes which contribute to locally experienced vulnerability. Depoliticization is not symptomatic of community-based approaches, which – in theory – offer greater scope for community actors to formulate their own discourses and promote their own agendas. Rather, it is a legacy of an enduring hierarchical system of governance, combined with a prevalent world-view of hazards and disasters which has become so entrenched as to require more than a shift in approach to overcome it.

## Conceptualizing vulnerability

Vulnerability is a socially constructed term that can be construed in different manners. By tying vulnerability to natural events in analysis, other facets of vulnerability and the links between them are neglected. Local people are subject to a far wider range of risks and stresses than those associated with natural hazards, and their coping strategies address these (Bhatt 1998; Burton *et al.* 1978).

I use the term 'event-centred' to describe a form of conceptual understanding of vulnerability that focuses upon (in the context of this study) typhoon-related manifestations of vulnerability. Event-centred understandings include the *root causes* of vulnerability manifested during events. Fundamentally, I argue that over-concentration upon event-centred vulnerability risks neglecting forms of underlying vulnerability that, unchecked, are likely to emerge as future causes of event or stress manifestations of vulnerability.

In practice, the tendency to focus upon hazard events has implications for community-based projects. When it comes to reducing vulnerability by use of a community-based approach, the ideas, knowledge and strategies of local people tend only to be considered in as far as they fit within the bounds of externally imposed programme or project definitions of vulnerability to disaster. Causes that can be directly linked to the effects of hazard events are incorporated in an event-centred conceptual framework. These causes of event-centred vulnerability can be traced to processes such as environmental degradation, building and infrastructural quality, population pressures and migration trends – the latter two contributing to greater numbers settling or reaping their livelihoods in marginal areas such as foreshore, riverside or erosion-damaged hillside locations. Coping strategies represent the ways in which people respond to event-centred vulnerability during, in the aftermath, or in preparation for events. Direct coping strategies such as stockpiling of basic commodities, developing warning systems and evacuation procedures, rebuilding homes and repairing infrastructure can be incorporated in an event-centred conceptual framework. In the Tigbao project an attempt was made to alleviate the effects of flash flooding by dredging and diverting the course of a stream that threatened, under typhoon conditions, to flow onto surrounding farmland. Event-centred vulnerability was thus alleviated, the *risk* of flood damage to this land having been significantly reduced. However, the underlying vulnerability of farmers to flood damage or other stresses was not reduced by this action.

One means of addressing underlying vulnerability, in the example cited above, might have been through the introduction of a livelihood scheme such as hog-raising or handicraft-making. Such a scheme would provide participants with an extra source of income to fall back upon in the event of a failing of their primary livelihood sources. Local strategies to cope with vulnerability are closely linked to livelihoods and are founded upon three core areas of capacity, security and opportunity. These provide a basis for more wide-reaching vulnerability reduction.

## Conclusion

This chapter has outlined an appropriate analytical framework for examining community-level vulnerability. The framework allows for the inclusion of a broad range of underlying factors and processes, agency, and different constructions of vulnerability, in analysis.

Community members operate on a day-to-day basis with much broader understandings of vulnerability that those encompassed in the project remit. Dominant, event-centred constructions of vulnerability restrict the types of activity that can be undertaken. Time and financial constraints, as well as organizational experience and procedures, serve to narrow the project remit. The segregation of different types of vulnerability (for instance, event- and livelihood-centred) is an artificial practice. In so far as such constructions of vulnerability are imposed from above (driven as they are by implementing organization and donor actors), I contend that they contradict fundamental principles of their own community-based approaches, and run contrary to underlying project objectives to increase local capacities to address vulnerability. In part, these findings are symptomatic of the introduction of a relatively new approach by a large and evolving organization, with a long history in disaster response. They are also attributed to the enduring influence of dominant approaches to the theory and practice of disaster management (Hewitt 1983, 1997).

From the community member perspective, livelihoods, not hazard events, are the primary source of vulnerability. Local manifestations of vulnerability are linked to such factors as land tenure patterns that limit access to land; processes of environmental degradation; lack of livelihood opportunities in the area; rising prices of agricultural inputs and basic commodities; and falling market values of local produce. Different manifestations of vulnerability are too strongly interlinked in the lives of most community members to separate neatly vulnerability to flooding or storm surge, from vulnerability to food shortage, or underemployment.

Theorists and practitioners are moving (albeit slowly) towards an integrated approach to cross-sectoral vulnerability reduction, which links disasters and sustainable livelihoods/development (see, for instance, Ashley and Carney 1999; Handmer and Dovers 1996; Twigg 2001). Traditional distinctions between disaster management and development are of practical significance to government bodies and non-governmental organizations such as PNRC on both sides of the disaster management–development divide. Such distinctions have less significance for local people, and particularly for those with low levels of livelihood security for whom 'the difference between normal life and what outsiders define as a crisis may be marginal' (Eade 1997: 166). 'Complex emergencies' (associated with situations of violent human conflict) are already treated as requiring particular attention to socio-political factors. Increasingly less are such situations addressed as purely humanitarian causes divorced from their socio-political context. I maintain that vulnerability to natural events is equally complex. Isolating vulnerability to events from the wider social context risks treating symptoms rather than causes. Isolating vulnerability to events from other manifestations of vulnerability predefines 'problems' and risks bypassing local priorities and realities.

## Notes

1 The *barangay* is the lowest political unit of the Philippine local government system, and corresponds approximately to 'village' in terms of size and cohesive quality. *Barangay* Tigbao has 1,453 inhabitants (Tigbao Barangay Council 1999).
2 The findings of this chapter are based upon my Ph.D. research, which has been sponsored by the International Federation of Red Cross and Red Crescent Societies, and supported by PNRC and project participants in Tigbao.
3 This position is supported by Twigg (2001).
4 At the time of migration, life in the uplands had been rendered dangerous by the combined activities of insurrectionists and the pursuing military.
5 See Korten (1980) on 'blueprint' versus 'learning process approaches'.
6 I employ the term 'depoliticize' in a similar sense to that described by Ferguson (1994) with reference to the impacts of development institutions and their discourses.

## References


Ashley, C. and Carney, D. (1999) *Sustainable Livelihoods: Lessons from Early Experience*, London: Department for International Development.

Bhatt, E.R. (1998) 'Women victims' view of urban and rural vulnerability', in J. Twigg and M.R. Bhatt (eds) *Understanding Vulnerability: South Asian Perspectives*, London: Intermediate Technology.

Blaikie, P., Cannon, T., Davis, I. and Wisner, B. (1994) *At Risk: Natural Hazards, People's Vulnerability, and Disasters*, London: Routledge.

Burton, I., Kates, R.W. and White, G.F. (1978) *The Environment as Hazard*, New York: Oxford University Press.

Cannon, T. (2000) 'Vulnerability analysis and disasters', in D.J. Parker (ed.) *Floods*, Volume I, London: Routledge.

Cernea, M. (ed.) (1991) *Putting People First: Sociological Variables in Rural Development* (2nd edition), Oxford: Oxford University Press.

Chambers, R. (1983) *Rural Development: Putting the First Last*, Harlow: Longman.

Chambers, R. (1993) *Challenging the Professions: Frontiers for Rural Development*, London: Intermediate Technology.

Chambers, R., Pacey, A. and Thrupp, L.A. (eds) (1989) *Farmer First: Farmer Innovation and Agricultural Research*, London: Intermediate Technology.

Eade, D. (1997) *Capacity-Building: An Approach to People-Centred Development*, Oxford: Oxfam.

Ferguson, J. (1994) *The Anti-Politics Machine: 'Development', Depoliticisation, and Bureaucratic Power in Lesotha*, Minneapolis: University of Minnesota Press.

Hall, N. (1997) 'Incorporating local level mitigation strategies into national and international disaster response', in J. Scobie (ed.) *Mitigating the Millennium: Proceedings of a Seminar on Community Participation and Impact Measurement in Disaster Preparedness and Mitigation Programmes*, Rugby: Intermediate Technology.

Handmer, J. and Dovers, S. (1996) 'A typology of resilience: rethinking institutions for sustainable development', *Industrial and Environmental Crisis Quarterly* 9 (4), 482–511.

Hewitt, K. (ed.) (1983) *Interpretations of Calamity*, London: Allen and Unwin.

Hewitt, K. (ed.) (1997) *Regions of Risk: A Geographical Introduction to Disasters*, Harlow: Longman.

Johnston, B.F. and Clark, W.C. (1982) *Redesigning Rural Development: A Strategic Perspective*, Baltimore, Md.: Johns Hopkins University Press.

Korten, D.C. (1980) 'Community organisation and rural development: a learning process approach', *Public Administration Review* 40 (2), 480–511.

Leonard, D.K. and Marshall, D. (1982) *Institutions of Rural Development for the Poor*, Berkeley: University of California Press.

Lewis, J. (1999) *Development in Disaster Prone Places: Studies of Vulnerability*, London: Intermediate Technology.

Luna, E.M. (2000) 'NGO Natural Disaster Mitigation and Preparedness: the Philippine Case Study', Available online: http://www.redcross.org.uk/ (accessed 28 March 2001).

Midgely, J. (1986) 'Community participation: history, concepts and controversies', in A. Hall, M. Hardiman, J. Midgely and D. Narine (eds) *Community Participation, Social Development and the State*, London: Methuen.

PNRC (1994) 'PNRC: strategies towards the 21st century: improving the situation of the most vulnerable', Unpublished, Manila: PNRC.

PNRC (1999) 'A report and proceedings of the ICDPP-DMS Integration Workshop July 12–16, 1999', Unpublished, Manila: PNRC.

Pretty, J.N. and Scoones, I. (1995) 'Institutionalizing adaptive planning and local-level concerns: looking to the future', in N. Nelson and S. Wright (eds) *Power and Participatory Development: Theory and Practice*, London: Intermediate Technology.

Shepherd, A. (1998) *Sustainable Rural Development*, London: Macmillan Press.

Southern Leyte Province (1997) 'Provincial Disaster Plan: 1998–1999', Unpublished, Maasin, Philippines: Provincial Government.

Tigbao Barangay Council (1999) '*Barangay* Development Plan', Unpublished, Libagon, Philippines: Municipal Government.

Twigg, J. (1998) 'Understanding vulnerability – an Introduction', in J. Twigg and M.R. Bhatt (eds) *Understanding Vulnerability: South Asian Perspectives*, London: Intermediate Technology.

Twigg, J. (2001) *Sustainable Livelihoods and Vulnerability to Disasters*, Working Paper 2/2001, Benfield Grieg Hazard Research Centre, Available online: http://www.bghrc.com (accessed 22 October 2001).

Uphoff, N. (1991) 'Fitting projects to people', in M. Cernea (ed.) *Putting People First: Sociological Variables in Rural Development*, Oxford: Oxford University Press.

Uphoff, N. (1992) *Learning from Gal Oya: Possibilities for Participatory Development and Post-Newtonian Social Science*, London: Cornell University Press.

Wade, R. (1986) 'Common property resource management in south Indian villages', in National Research Council on Common Property Resource Management, *Proceedings of the Conference on Common Property Resource Management*, London: National Academy Press.

Watts, M. (1983) 'On the poverty of theory: natural hazards research in context', in K. Hewitt (ed.) *Interpretations of Calamity*, London: Allen and Unwin.

White, G.F. (ed.) (1974) *Natural Hazards: Local, National, Global*, London: Oxford University Press.

Winchester, P. (1992) *Power, Choice and Vulnerability: A Case Study of Disaster Mismanagement in South India, 1977–1988*, London: James and James Science.

# 12 Risk regime change and political entrepreneurship

## River management in the Netherlands and Bangladesh

*Jeroen Warner*

### Introduction

Over the past two decades the approach to river management has seen great change. Concrete structures, straight channels and top-down management have given way to greener, more complex engineering and participatory decision-making regimes. This chapter examines how individual political entrepreneurship within the state might trigger such change. Entrepreneurial actors seek to bring innovative ideas for managing natural resources onto the policy agenda for dealing with risk. In so doing they introduce elements of a new paradigm into the water management regime. Different literatures are connected with each other to develop an analytical framework for understanding innovation, which is applied to two cases – one in the North (the Netherlands) and one in the South (Bangladesh).

### Uncertainty and complexity in public management

The past few decades have seen an uneasy but necessary reassessment of the 'engineerability' of society. Social engineering failed to solve problems in the societal domain, while civil engineering did not put an end to risks such as flooding and water scarcity. The intractability of many social development issues coincides with a feeling in some quarters that the governability of society itself is at risk (Kaplan 1994). But faced with a complex problem, we can at least try to increase the solution space. For issues that are both complex, dynamic and diverse, Kooiman (1993) suggests an approach in which governments abandon their going-it-alone approach and share governance tasks and capacities with the private and civil society sectors, on the basis that neither top-down, bottom-up nor horizontal (co-governance) approaches alone can work. A multilevel, multi-actor, multifaceted, multi-instrumental and multi-resource based (Bressers and Kuks 2001) mode of governance should improve the relation between stress and coping capacities; that is, the *governability* (Kooiman 1993; Green and Warner 1999) of today's complex policy issues.

This is in line with governance developments in general, both in Western Europe and elsewhere, where overloaded governments are at pains to share some responsibilities in water management. The mixed results of privatization turned attention to a long-neglected civil society. Today's governance is like network management, in which top-down, lateral and spontaneous bottom-up actions together decide how an issue is addressed (Kooiman 1993). This means that power is not exercised 'over' other actors, but 'with' them. For individual actors, recognition of a common problem is not a sufficient condition for joining a regime; choosing to join is likely to be influenced by the attractiveness of proposed solutions, even if at times the problem itself is of little concern to the actors involved (Gupta *et al.* 1995).

The present chapter seeks to make sense of regime change in the water management sector. Unlike many other sectors, water management has a high potential for conflict, being characterized by complex interdependence because of the diversity of its uses and users, and because it involves life-and-death issues such as scarcity and flooding. It also involves great uncertainty – water flows are notoriously hard to model, and floods remain hard to predict. Floods are now predicted to become more variable and damaging due to climate change (Hisschemöller and Olsthoorn 1999). This makes it even more necessary to be prepared for the inevitability of catastrophic flood events, accepting failure and accepting a (greater) degree of uncertainty. Involving diverse stakeholders in water management increases the importance of the human factor in a sector that has long been technology-dominated. This complexity may not just drive up the cost; also, from the point of view of the initiator, it may increase risks to the successful completion of the project itself. The acceptance of complexity and uncertainty has had a significant impact on the river management 'regimes'; that is, the sets of norms, rules and procedures by which an issue area is governed. We will now briefly look into the debate on regime change.

## *Regimes and innovation: the theoretical challenge*

Regime theory emerged as a way of understanding co-operation and structure in international relations. As international water conflict started to attract interest (Gleick 1993; Bulloch and Darwish 1993), students of public administration and law started applying regime theory to the regulation of conflict over shared water resources – for example, the regulation of water bodies such as the Mediterranean (Haas 1992) and the Rhine (Dieperink 1997). Other fields are now applying regime theory at different geographical scales, such as regional resource regimes (Arnesen 1997) and urban regimes (Jewson and McGregor 1997; Stoker 1995). Here we focus on initiatives from within the public sector, and look into regime change and the role of political entrepreneurship in driving that change. This focus implies an epistemological choice between different approaches to regime theory in favour of an actor approach (see Box 12.1) as an alternative to the still-dominant, legal-institutional approach, which has a tendency to look at rules without attaching them to roles and action.

**Box 12.1 Three approaches to regime theory**

- The predominant approach focuses on the dominant rules, decision-making procedures, norms and values in an issue area (Krasner 1982). A rule-based orientation, however, circumvents the question of who acts on what (agency dilemma): it is not actor-oriented, but centres around rules.
- In neo-realist varieties of regime theory, actors have *roles*. An actor-oriented school looks at interactions in actor networks where actors find themselves lead actors, veto actors or facilitators (Porter and Welsh-Brown 1995).
- In a cognitive approach regimes purport to reduce 'noise' in providing a clearing house for information, enabling the participating actors to learn. New functional knowledge may lead to evolutionary change, changing rules and procedures as the regime 'learns', or revolutionary change, generating new principles and norms, and is associated with power shifts. According to Haas (1992), this convergence process leads to the formation of 'epistemic communities' converging on a body of accepted scientific procedure and evidence.

Sources: Haggard and Simmons (1987), Hasenclever *et al.* (1997)

Actor-oriented approaches, in this perspective, provide an entry-point for a 'revolutionizing' approach. The possibility of a revolutionizing approach is important in light of the challenge Susan Strange made quite early on in the regime debate. Strange's problem with regime theory is that it is inherently conservative – it is oriented towards maintaining the status quo (Strange 1982). Regimes may hide structural inequalities, or increasingly find themselves out of touch with changing social demands, calling forth a competing 'counter-hegemonic regime' as a result. Group-think, complacency and self-selection may creep in – after all, regimes not only tie actors together, they also keep actors (and their knowledge) out (Bühl 1995). New alternatives and paradigms may be kept outside of accepted practice until a window of opportunity opens. Solutions may even lie in waiting until suitable problems appear on the horizon to which they can attach themselves (Cohen *et al.* 1972).

An agency-oriented view would argue that a window can be opened actively by political entrepreneurs (Young 1994; Moravcsik 1998–1999) who come up with a bold idea and manage to build a constituency for it. Entrepreneurs are change agents, who are able to produce something out of nothing,

*Table 12.1* Cultural theory of risk

| *Group* | *Hierarchists* | *Individualists* | *Egalitarians* | *Fatalists* |
|---|---|---|---|---|
| View of risk | Risk managing | Risk taking | Risk avoiding | Risk expecting |
| View of nature | Nature perverse/ tolerant | Nature benign | Nature ephemeral | Nature capricious |

Sources: After Holling (1979), Douglas and Wildavsky (1982), Thompson *et al.* (1990).

often to their individual personal and/or professional risk. The cultural theory of risk identifies four personality types (hierarchists, individualists, egalitarians and fatalists) and predicts that it is individualists who are most likely to accept and exploit risk (see Table 12.1) (Thompson *et al.* 1990). Each personality type yields a preferred state of organization which has subsequently been linked to four 'myths of nature' necessary as assumptions for their preferred way of life. The fundamental implication is that different cosmologies bring in different knowledges relating to different degrees of risk acceptance and values where the state of the world is unclear.

In water management the primary perspective is technocratic, dominated by experts. However, while it is tempting to box in engineers with the control-oriented hierarchists, there are those who enjoy the challenge of being in the avant-garde, and have displayed individualist behaviour. We can expect them to be helpful in locating and implementing opportunities for adaptation (Kooiman *et al.* 1999). Such entrepreneurial experts may be pivotal in forming transdepartmental (Stoker 1995) or transboundary (Haas 1992) coalitions for changing the regime.

While entrepreneurialism has an attractive ring to it, from the viewpoint of the existing regime attempts to bend or change the rules may be seen to involve a degree of deviance, a form of negative behaviour to be disciplined. However, if nobody broke any rules and got away with it, there would be little incentive for innovation. This chapter suggests that regime change may involve deviance, in the sense of bending or breaking rules. There are two levels of 'rule' in an institutional regime: first-order rules are direct injunctions born out of second-order 'meta'-rules, determining how first-order rules can be created, modified and rescinded (Kratochwil and Ruggie 1986).

Tolerance of deviance can either indicate a lack of alternative methods for forcing adjustments, or be a result of an extremely permissive, flexible regime. Regime flexibility can be an expression of resilience to shocks, and therefore of regime strength. On the other hand, if certain actors can afford to get away with just about anything and secondary rules are very weak (at least for them), the political regime could be described as weak. Thus, first-order non-compliance only indicates regime weakness if second-order compliance is weak as well.

Management scientist Gary Hamel (2000) uses the metaphor of a revolution to discuss rule breaking and innovation. He presents a seven-step platform for starting an insurrection within a company. At each stage the intensity of change potential increases. In order these stages are: (1) establish a point of view, (2) write a manifesto, (3) create a coalition of like-minded rebels, (4) pick your targets in senior management, (5) co-opt and neutralize your targets, (6) find a translator to advocate your views, (7) when pushing for change aim to win small, win early but win often.

## Innovation in water management: two cases

The building blocks introduced above are applied below to attempts at regime change in the water management systems of the Netherlands and Bangladesh. In both cases, the catalyst for change was a mid-ranking official and change was preceded by a period of impasse between so-called blue lobbyists (control-oriented water engineers) and green lobbyists (ecologists and participationists). In the Dutch case, amidst a thriving national economy, change was secured leading to the introduction of a new form of natural resource management in riverine wetlands. In Bangladesh, in a chronically dependent economy, change-agents sought to reform operation and maintenance (mal)practices by increasing local participation in the management of flood-prone polders. First, it is worth sketching the changing international water management paradigm within which innovation took place.

According to Green and Warner (1999), we are now in the fourth wave of approaches to flood hazard management. After a first, pre-modern, 'unmanaged' episode in which people adjusted their lives to flooding, a second wave followed, fitting nature to human needs and making rivers more efficient at carrying away flood waters. The approach of the third wave was almost exactly the converse: human behaviour should be adjusted through institutional actions to fit the behaviour of the river. It tries to arrest further development of the flood plain. Both the second and third waves are characterized by a static world-view, an obsession with equilibrium. The emerging fourth-wave paradigm allows nonlinearity and a fair degree of uncertainty (Geldof 1994). The new model questions the effectiveness of different, isolated policy options, favouring a mixture of policy instruments. It is characterized by catchment-based approaches preferring multi-functional solutions for different parts of a single catchment.

Neither the Netherlands nor Bangladesh have regime contexts conducive to internal innovation, if for different reasons.

In the Netherlands, the historical battle with water has shaped political culture. On the one hand, the struggle against water gave rise to the Netherlands' oldest democratic institutions for collective action with a history dating back to the thirteenth century. In these *waterschappen*, stakeholders pooled their resources to build dikes and embankments. On the other hand, the need to keep everyone on board gave rise to a consensus-oriented culture and a sluggish rate of policy change. The consensual orientation does not make for an innovative culture (Jacobs *et al.* 1989). Visionary ideas and eccentricity are not necessarily welcomed. On the other hand, the Low Countries have a history of bold moves when faced with acute danger that has legitimized breaks with consensual politics and released resources for change. Frissen (1998) summarizes this legacy as a tendency to ignore ideological gaps for the greater good, habituation to a functional (problem-oriented) rather than territorial form of management, a technocratcic mode of operation when faced with existential issues and a belief that Holland can be engineered.

In Bangladesh, inertia must be seen in the context of a relatively young, violent and undemocratic polity (Ahsan 2000). Since soon after independence in 1971, partisan political competition and ethnic violence has obstructed and disrupted the functioning of the public sector (Kochanek 1993). Bangladesh is a low-trust society in Fukuyama's terms (1995). Kochanek (1993) also identifies the deeply ingrained patron–client system mediating access to and influence on the highly centralized political system. The Local Government Act of 1996 has devolved very few powers to lower levels. At the time of President Ershad, when the Flood Action Plan was initiated, each and every single development project needed the President's signature. Because of the internal political stalemate any impulses for change tend to come from the outside. Bangladesh is highly adaptive to external demands for institutional change with a view of ensuring the continued inflow of donor money, but Bangladesh's bureaucratic culture does not lend itself well to adopting new ideas from civil society.

This is different in the Netherlands, where change through 'reflexive modernization' can be seen in the Waterways Agency (Rijkswaterstaat) being forced to work with the Environmental Department (VROM) and incorporate participation and nature development, as reflected in the Fifth Memorandum on Public Planning, (http://www.vrom.nl/pagina.html?id=1&goto=3410&site=www.vijfdenota.nl). Another structural innovation, effectuated in the 1980s, saw public officials being given a fair degree of freedom to participate in unofficial interdepartmental working groups, which have led to innovative documents like the Plan Ooievaar and, more recently, Breinstroom, on multiple uses of space in basin management.

### *Plan Ooievaar (the Netherlands)*

The Dutch case starts with an institutional power struggle over dominance in water management. The *waterschappen* (water management boards) are typically Dutch civil society organizations. Over the course of time, constant bickering between adjacent water management boards at times led to serious conflict and obstruction. Rijkswaterstaat (RWS), a public agency created at the turn of the nineteenth century, was intended to provide a unitary national body. Flood emergencies provided unique opportunities for RWS to wrest power from the water management boards. When the 1953 flood 'proved' that the water management boards had neglected the state of the sea defences, RWS stepped in with a giant Delta Plan (Bijker 1995).

In the 1970s the closing piece of the Delta Works, the Oosterscheldt Dam, became a *cause célèbre* for a growing NGO movement that claimed water projects had negatively impacted on the environment. Within RWS a generational conflict was raging, with more and higher dikes advanced by the older generation and local flood proofing and adaptation championed by younger enegineers. The 'Plan Ooievaar' arrived at a time when a window for greening water management coincided with a change in dominant thinking on nature management. It had become clear in nature management quarters that if you stop dumping things in rivers, natural regeneration can take place with surprising speed. This 'nature development' approach provided an exciting alternative to the conservatory approach allowing the exploitation of ecological dynamics to bring about environmental improvement.

A window of opportunity for change opened when the Transport and Waterways Minister, Smit-Kroes, invited plans for innovation targeting senior management (Hamel's fourth step). A transdepartmental coalition between RWS and what is now known as LNV (the Agriculture, Nature and Fisheries Department) emerged as Dick de Bruin, a mid-ranking officer at LNV, saw a possibility to promote his agenda of nature development, which broke both with the engineering approach and nature conservation approach.

De Bruin teamed up with civil engineers in Rijkswaterstaat to produce the winning entry. The report, known as the Stork Plan (de Bruin *et al.* 1987) served as a manifesto (the second of Hamel's steps) for nature development. It abandoned the conservation and target-oriented approach for a process-oriented approach in which nature development was promoted rather than steered. It also abandoned the idea that agriculture should support natural development, and linked nature with other functions such as water supply and leisure. In so doing, the Ooievaar team defied conventional wisdom within the Agricultural Ministry and latched onto a new, more innovative mood within the Waterways Department. De Bruin strategically teamed up with civil servants from a different ministry and used these contacts to address the Minister directly. He knew the Minister was nearing the end of her term and had relatively little to lose from a daring move. His boldness worked wonders; that same day the Minister announced she wanted to promote the nature development approach. This amounted to a benign, enterpreneurial form of 'deviance'.

### *Aftermath*

The idea of nature development took root, leading to a report on the river Maas management (Stroming 1991). This report helped shape an ambitious plan for deepening and widening the river Maas, with interventions that were intended to facilitate spontaneous nature development rather than forcing defined outcomes. The *waterschappen* and Rijkswaterstaat were won over and the environmental NGO Natuurmonumenten was ready to participate in the consortium of public and private parties in what was basically a risk-bearing investment given unclear costs and benefits. In 1993 and 1995 riverine floods tilted the balance between blue and green concerns. A Delta Plan was urged, this time for securing rivers from flooding. However, the concept of accepting uncertainty with a view to nature development keeps popping up in policy documents and visions (most notably WL Delft 1998). In that light, it seems that the innovation has been a qualified success in progressing the regime.

### *Participation in the SMB polder (Bangladesh)*

Similar to the Dutch green and blue world-views, flood policy in Bangladesh and its predecessor, East Pakistan, has oscillated between two concepts of control: a strong belief in flood control and a more cautious 'living with the floods' approach (cf. Wood and Faaland 1995). In more recent years, a corresponding debate has emerged about whether development should be surface-water or groundwater based. In contrast to the Netherlands, there has always been a strong overlay of external ideas on flood policy in Bangladesh. Paralysed by its internal stalemates and happy to rely on external aid, Bangladesh has moved with the waves of prevailing international flood policy ideologies. This can be seen in the mid-twentieth century, when, coloured by the Green Revolution, large engineering solutions were in vogue, justified by the agricultural gains to be made. For example, the 1964 Master Plan included 58 engineering projects. But many projects became marred by deficiencies in the operation and maintenance of the structure. The World Bank recommended smaller water works in future, and in response to this external influence a national programme to promote low-lift pumps and small-scale irrigation was introduced.

In 1987 and 1988 Bangladesh experienced two devastating floods. The 1988 flood put 60 per cent of the country under water for two weeks, damaging 7.2 million homes, and, unusually, affecting the well-to-do, including the home of the then President, General Ershad, as well as the army cantonment. Among the many mishaps striking Bangladesh and other developing countries every year, the reason why the issue reached the top of the international agenda seems fortuitous. Mme Danielle Mitterrand happened to visit Dhaka and Tangail and, shocked, raised the issue with her husband, then president Mitterrand in France, who was eager to raise France's profile in the world as a benefactor. This helped raise the political profile of flood aid to Bangladesh, and Mr Jacques Attali was called to lead a team of 30 experts to draw up a 'permanent solution' to flooding in Bangladesh.

President Mitterrand promoted the bold idea of putting an end to floods in Bangladesh at the G7 conference in Paris in July 1989. Eleven guiding principles were drawn up, which combined nods at non-structural works and participation with river training and channelling. On the whole, though, little attention was given to non-structural measures. A Delta Plan-like

Flood Control and Drainage (FCD) plan was the starting point for the Flood Action Plan (FAP). While straightforward in engineering terms, its operation brought with it a host of socio-economic, environmental and institutional issues which unexpectedly politicized the project. Soon, any pretence of flood protection had to be abandoned in favour of controlled monsoon flooding.

After the return to multi-party democracy a government-appointed Task Force in 1991 recommended a moratorium on structural works, including the FAP. The attraction of a massive inflow of aid was great however, and the task force report went unheeded. However, the report bolstered a latent NGO concern in fighting the principles of the FAP. Among a host of other concerns (Haggart 1994) the Plan was criticized because it seemed externally imposed without paying due heed to what local people wanted. FAP-20 was oriented at improving the efficiency of monsoon farming, against a strong local preference for drainage and improving groundwater-irrigated winter (rabi) crops (Adnan 1992). NGOs helped organize mass protest in Tangail and Dhaka and at a FAP conference in the European Parliament. This made the donors put pressure on the Government of Bangladesh (GoB) to press for reform. In 1994, the GoB approved guidelines for participation in water management prepared by the Ministry of Water Resources (http://www.unescap.org/drpad/publication/integra/modalities/bangladesh/4bl03e03.htm). In practice, however, very little was happening due to resistance to change on the part of the Bangladesh Water Development Board (BWDB). Many BWDB officials clearly did not believe in participation; in my interviews several professed to have low regard for people's understanding of water systems.

Against this backdrop, the late 1990s saw an initiative from within BWDB. M.A. Quassem, a mid-ranking BWDB engineer who had studied social science as well as engineering, saw an opportunity to try participation in his area, the Sreenagar–Mawa–Bhagyakul (SMB) Project, 50 km from the capital of Dhaka, serving 80,000 people. The project is a flood control and drainage project, but in fact has been turned into an irrigation project by private individuals as all the land is irrigated in the dry season (winter) to ensure the successful harvest of transplanted *aman* (a popular type of rice).

Until this initiative there had been five or six breaches and 'public cuts' every year, physical evidence of people's unhappiness with the way drainage was arranged and timed. Dr Quassem, born from a respected family in this area, commanded enough influence for people to try something new. Quassem's idea was to form embankment protection committees in eight unions (an administrative unit comprising a number of villages) crossed by the embankment. Farmers, fishermen, local leaders and women (as a separate category) were invited to create an example of operation and maintenance activities involving local stakeholders. A complex arrangement was drawn up to co-opt the BWDB and local elites in the scheme, which gave

BWDB field staff a new role in the management as technical advisers and facilitators. A two-tier system was devised, at union and polder level, with BWDB officers serving as *ex officio* members (Quassem, n.d.).

Influential people were also involved to handle the powerful *mastans* who threaten ordinary people with violence. Weak places in embankments were identified and repaired, some by BWDB as a regular work but others by local people at a fraction of the cost and roughly the same quality. These savings were possible because on formal projects a huge percentage is siphoned off as pilferage. And indeed, the next year there were no breaches or cuts. However, the regime bit back: a proposed allocation was suddenly cancelled – not just to rehabilitate after the 1998 floods but to repair the breaches. There was to be a seminar on the SMB experience, which fell through. After this, little was heard about the SMB initiative, while its initiator moved on to greater things as Chairman of the Water Resource Planning Organization (WARPO).

### *Aftermath*

The apparent failure of the initiative sheds an interesting light on the sectoral reform initiatives that are currently underway in Bangladesh. There is a strong sense that BWDB does not really want to be involved in participation, or even in operation and maintenance. There has been a recent and novel twist with the new BWDB Water Sector Improvement Plan currently prepared by a $70 million World Bank Bangladesh–Water Sector Improvement Project (see http://wbln0018.worldbank-bangladesh.org/lo%20web%20sites/bangladesh%20Web.nsf/0704a4348e105b2e46256672002 3975f/826c69abee218cbc4625671b0028708b?OpenDocument). This plan seeks to hand over the operation and maintenance of water projects to the community without providing money – it is expected to be raised from local taxes and collection. Under the banner of participation, the government is working on new guidelines to turn over the operation and management of flood works in areas up to 5,000 ha to local communities, while larger embankments would be managed by the Local Government Engineering Department, with the Bangladesh Water Development Board and community providing inputs (Water Sector Improvement Plan). This seems a case of offloading rather than seriously involving the grassroots in operating and maintaining flood defences!

Superficially, extremely exciting changes seem to have taken place in Bangladesh. The National Water Management Plan (WARPO 2000) seeks an integrated approach, including environmental and social concerns. NGOs were co-opted in the board supervising the participatory exercise to tease out people's preferences. However, innovations seem to have been mainly externally driven. The Bangladeshi NWMP plan was drawn up and implemented by the British consultant Halcrow, and there is a strong impression that alternatives presented to the people during public consultations were

*Table 12.2* Innovation trajectories

| *Hamel's seven steps* | *Netherlands* | *Bangladesh* |
|---|---|---|
| 1 Establish a point of view | X | X |
| 2 Write a manifesto | X | |
| 3 Create a coalition of like-minded rebels | X | X |
| 4 Pick your targets in senior management | X | |
| 5 Co-opt and neutralize | X | X(?) |
| 6 Find a translator | | X |
| 7 Win small, win early, win often | ? | |

strongly biased to hi-tech solutions. As ever, support from BWDB for participatory reform appears nominal.

### *Hamel in practice*

Returning to the seven steps advanced by Gary Hamel, Table 12.2 tentatively presents the innovation trajectories followed in the Bangladesh and Netherlands. In both cases, opportunity was actively pursued by enterprising individuals with a clear point of view (1) enlisting like-minded rebels (3). In the Dutch case, an attractive manifesto (2), backed up by an award, worked wonders. Material backing and commitment, however, needs to back up these efforts, which is where difficulties commence. In Bangladesh, the BWDB officer clearly did not successfully win over senior management (4), while the Dutch public servant did. In both cases they sought to co-opt (5) the powers that be, and to act as a translator (6) between the middle rank and the top of the organization. A failure to achieve the seventh step, winning small, early and often was observed in Bangladesh, but also marred the Dutch case despite early optimism. An early (big) win was followed by several small victories incorporating nature development in other projects. However, there is always a danger that when the money runs out, the momentum for nature development will run out too.

## Conclusion

This chapter has provided two examples of entrepreneurship within public organizations seeking to change the way water hazards are mitigated. In the Dutch case, a flexible regime was revolutionized by an entrepreneurial group making the most of a window of opportunity. In the Bangladeshi case, the same kind of entrepreneurial deviance was stifled by a rigid regime. These studies illustrate a framework for understanding regime change at the state–society interface, in this case taking place within bureaucracies. It has been argued that a network management view helps explain how initiatives within the network can change the outlook of the whole network, as happened in the Netherlands, by seizing windows of opportunity.

It should be noted that the cases discussed here are attempts on the part of experts within the public sector to revolutionize water management. It would be interesting to see whether regime revolutions might also be initiated by non-expert entrepreneurs, or political entrepreneurs from civil society (cf. Haufler 1993). The attempts at change described in this contribution were set in the context of an international flood management paradigm, which is moving towards greater complexity and user involvement.

The Dutch innovators were hampered by a hyperconsensual culture, but found an attractive way of introducing an innovative concept for nature development, promoted by a contest (*prijsvraag*), at a time when senior management was at a crossroads. In Bangladesh, opportunities for change at the same kind of crossroads tend to be either blithely ignored or forced to change by external donors. The political sector was caught up in disruptive politics, while the bureaucratic sector had little incentive to change its entrenched ways. This stood in the way of any long-term vision and killed innovation, unless imposed from the outside.

In both countries, change-oriented initiatives were always at risk due to a lack of funding. Neither nature development nor public participation seemed immediately rewarding to the public sector from a cost–benefit perspective. However, at base it seems to be differences in institutional organization and political culture between Bangladesh and the Netherlands that explain why the Dutch innovation was reasonably successful while the Bangladeshi innovation was not. The Dutch system, despite its treacly decision-making process, incorporates sufficient flexibility to allow interdepartmental alliances for change. In Bangladesh, both institutional and political rigidities cause stalemates that tend to move only after outside pressure has been applied.

## Acknowledgements

The research underlying this chapter was carried out in the context of a Ph.D. project funded by the Flood Hazard Research Centre, Middlesex University, Queensway, Enfield, Middlesex, United Kingdom EN3 4SF. The author thanks Dick de Bruin and M.A. Quassem for sharing their experiences with me, Lennard Roubos of ICCO (the InterChurch Organization for Development Co-operation at Zeist, Netherlands) for coining the phrase 'reverse co-optation' in the course of our discussions, and Mark Pelling and Katrina Allen for their helpful comments on an earlier version of this contribution, the responsibility for which of course remains all mine.

## References

Adnan, S. (1992) *People's Participation, NGOs, and the Flood Action Plan: An Independent Review*, Dhaka: Research and Advisory Services.

Ahsan, S.M. (2000) 'Governance and corruption in an incomplete democracy: the Bangladesh case', Paper presented at the Sixth Workshop of the European Network of Bangladesh Studies, University of Oslo, 14–16 May, Oslo.

Arnesen, T. (1997) 'A method for analysis of ecological communication in regional management complexes (REMA)', Paper presented at the Athens International Conference Urban Planning and Informatics to Planning in an Era of Transition, 22–24 October, Athens, Greece, Available online: http://www.ostforsk.no/per/Tor.Arnesen/publikasjoner/REMA.htm (accessed June 2002).

Bijker, W.E. (1995) 'Sociohistorical technology studies', in S. Jasanoff, G.E. Markle, J.C. Petersen and T. Pinch (eds) *Handbook of Science and Technology Studies*, Thousand Oaks, Calif.: Sage.

Bressers, H. and Kuks, S. (2001) 'Governance patroness als verbreding van het beleidsbegrip', *Beleidswetenschap* 15 (1), 76–103.

de Bruin, D., Hamhuis, D., van Nieuwenhuijzen, L., Overmars, W., Sijmons D. and Vera, F. (1987) *Plan Ooievaar, de toekomst van het rivierengebied*; see also Plan Ooievaar, Available online: <http://www.arknature.nl/ark-lannen/rivier/rijntakken/ooievaar.asp> (accessed June 2002).

Bühl, W.L. (1995) 'Internationale regime und europäische integration', *Zeitschrift für Politik* 42 (2), 122–148.

Bulloch, J. and Darwish, A. (1993) *Water Wars: Coming Conflicts in the Middle East*, London: Gollancz.

Bureau Stroming (1991) *Toekomst voor een grindrivier*, Nijmegen.

Cohen, M.D., March, J.G. and Olson, J.P. (1972) 'A garbage can model of organizational choice', *Adminstrative Science Quarterly* 17 (5), 1–25.

Dieperink, C. (1997) 'International regime development: lessons from the Rhine catchment area', *TDRI Quarterly Review*, Available online: <www.info.tdri.or.th/library/quarterly/text/s97_4.htm> (accessed June 2002).

Douglas, M. and Wildavsky, A. (1982) *Risk and Culture. An Essay on the Selection of Technological and Environmental Dangers*, Berkeley: University of California Press.

Frissen, P.H.A. (1998) 'De zoete smaak van stroop. Een postmoderne visie op de consensusdemocratie', in F. Hendriks & Th. Toonen, *Schikken en Plooien*, Assen: Van Gorcum.

Fukuyama, F. (1995) *Trust: The Social Virtues and the Creation of Prosperity*, New York: Free Press.

Geldof, G.D. (1994) *Adaptief waterbeheer*, Deventer: Tauw.

Gleick, P.H. (1993) *Water in Crisis, A Guide to the World's Fresh Water Resources*, Oxford: Oxford University Press.

Green, C. and Warner, J. (1999) 'Flood management: towards a new paradigm', Paper presented at the Stockholm Water Symposium, Workshop 4, Stockholm, 9–12 August.

Gupta, J., van der Wurff, R. and Junne, G. (1995) *International Policies to Address the Greenhouse Effect. An Evaluation of International Mechanisms to Encourage Developing Country Participation in Global Greenhouse Gas Control Strategies, Especially Through the Formulation of National Programmes*, Amsterdam: Free University, Institute for Environmental Studies, Amsterdam.

Haas, P.M. (1992) 'Epistemic communities and international policy coordination', *International Organization* 46 (1), 1–35.

Haggard, S. and Simmons, B.A. (1987) 'Theories of international regimes', *International Organization* 41 (Summer), 491–517.

Haggart, K. (ed.) (1994) *Rivers of Life*, Dhaka: BCAS/PANOS.

Hamel, G. (2000) *Leading the Revolution*, Harvard: Harvard Business School Press.

Hasenclever, C., Mayer, P. and Rittberger, V. (1997) *International Regimes*, Cambridge: Cambridge University Press.

Haufler, V. (1993) 'Crossing the boundary between public and private: international regimes and non-state actors', in V. Rittberger and P. Mayer (eds) *Regime Theory and International Relations*, Oxford: Oxford University Press.

Hisschemöller, M. and Olsthoorn, A.A. (1999) 'Identifying barriers and opportunities for policy responses to changing climatic risks', in T.E. Downing, A.A. Olsthoorn and R.S.J. Tol (eds) *Climate, Change and Risk*, London: Routledge.

Holling, C.S. (1979) 'Myths of ecological stability', in G. Smart and W. Standbury (eds) *Studies in Crisis Management*, Montreal: Butterworth.

Jacobs, D., Lynch, J., Wilderom, C. and Zegveld, W. (1989) 'The Netherlands as a national system of innovation. A confrontation with Michael Porter's approach', Contribution to the Ninth Annual Strategic Management Society Conference, San Francisco 11–14 October.

Jewson, N. and McGregor, S. (1997) *Transforming Cities*, London: Routledge.

Kaplan, R.D. (1994) 'The coming anarchy', *Atlantic Monthly* 273 (2), 44–76.

Kochanek, S. (1993) *Patron–Client Politics and Business in Bangladesh*, New Delhi: Sage.

Kooiman, J. (ed.) (1993) *Modern Governance*, London: Sage.

Kooiman, J., van Vliet, M. and Jentoft S. (eds) (1999) *Creative Governance. Opportunities for Fisheries in Europe*, Aldershot: Ashgate.

Krasner, S.D. (1982) 'Regimes and the limits of realism: regimes as autonomous variables', *International Organization* 36 (2), 490–517.

Kratochwil, F. and Ruggie, J.G. (1986) 'International organization: a state of the art on an art of the state', *International Organization* 40 (4), 753–775.

Moravcsik, A. (1989–1999) 'A new statecraft? Supranational entrepreneurs and international cooperation, papers presented, 1998–1999, IR colloqium, Institute of International Studies, Berkeley, California', Available online: <http://globetrotter.berkeley.edu/irforum/papers/morav-int.html> (accessed June 2002).

Porter, G. and Welsh-Brown, J. (1995) *Global Environmental Politics*, Boulder, Colo.: Westview Press.

Quassem, M.A. (n.d.) *Water Institutions: Bangladesh Experience*, Available online: http://lnweb18.worldbank.org/ESSD/essdext.nsf/18DocByUnid/06SE9AOC6403095B85256BEA0046E233/$FILE/Quassem.pdf/

Stoker, G. (1995) 'Regime theory and urban politics', in D. Judge, G. Stoker and H. Wolman (eds) *Theories of Urban Politics*, London/Thousand Oaks, Calif.: Sage.

Strange, S. (1982) 'Cave! hic dragones: a critique of regime analysis', *International Organization* 36 (2), 337–354.

Thompson, M., Ellis, R. and Wildavksy, A. (1990) *Cultural Theory*, Boulder, Colo.: Westview Press.

Water Resource Planning Organisation (WARPO) (2000) *National Water Management Plan*, Available online: http://www.warpo.org/dNWMP/dNWMP.htm (accessed May 2002).

WL Delft (1998) *De Rijn op termijn*, Delft.

Wood, G.D. and Faaland, J. (1995) *Flood and Water Management: Towards a Public Debate*, Report by the Independent Review Mission on the Flood Action Plan in Bangladesh, Dhaka: UNDP.

Young, O. (1994) *International Governance, Protecting the Environment in a Stateless Society*, London: Cornell University Press.

# 13 Disaster as manifestation of unresolved development challenges

## The Marmara earthquake, Turkey

*Alpaslan Özerdem*

## Introduction

On 17 August 1999, at 03.02, 41,000 square kilometres between Bolu and Istanbul was struck by an earthquake registering 7.4 on the Richter scale. This is home to 23 per cent of the country's population and accounts for 34.7 per cent of GNP. The epicentre was located in Gölcük, the country's most important naval base, in the province of Kocaeli, 90 kilometres east of Istanbul (see Figure 13.1). The official death toll was 17,127, with an additional 43,953 hospitalized injuries. Estimated monetary losses were between US$9 and 13 billion, with direct physical damage being: industrial (US$2 bn), buildings (US$5 bn) and infrastructure (US$1.4 bn). Systemic economic losses doubled these figures, with a decrease in 1 per cent of GNP growth. An additional US$3.5 bn was expended on rehousing (DPT 1999). The high level of damage inflicted on the housing stock and commercial buildings of the area can clearly be seen in Table 13.1.

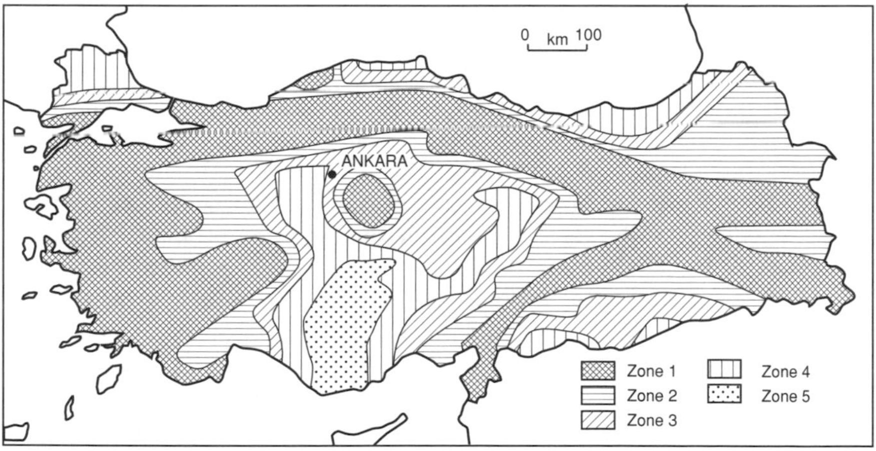


*Figure 13.1* Earthquake zoning map for Turkey.

Source: Özmen *et al.* (1997).

*Table 13.1* Building damage, Marmara earthquake

| *Number of damaged buildings* | | | | | |
|---|---|---|---|---|---|
| *Heavy to collapsed* | | *Medium* | | *Light* | |
| *Households* | *Commercial* | *Households* | *Commercial* | *Households* | *Commercial* |
| 66,441 | 10,901 | 67,242 | 9,927 | 80,160 | 9,712 |

Source: Government Crisis Center (1999).

*Table 13.2* Comparative evaluation of the Marmara earthquake

| *Place* | *Date* | *Magnitude on Richter scale* | *Number of deaths* | *Number of collapsed and heavily damaged buildings* |
|---|---|---|---|---|
| Gediz | 28.03.1970 | 7.2 | 1,086 | 9,542 |
| Burdur | 12.05.1971 | 6.2 | 57 | 1,389 |
| Bingöl | 22.05.1971 | 6.7 | 878 | 5,617 |
| Lice | 06.09.1975 | 6.9 | 2,385 | 8,149 |
| Çaldiran | 24.11.1976 | 7.2 | 3,840 | 9,552 |
| Erzurum | 30.10.1983 | 6.8 | 1,155 | 3,241 |
| Kars | 07.12.1988 | 6.9 | 4 | 546 |
| Erzincan | 13.03.1992 | 6.8 | 653 | 6,702 |
| Dinar | 01.10.1995 | 6.0 | 94 | 4,909 |
| Adana | 27.06.1998 | 6.3 | 146 | 4,000 |
| Marmara | 17.08.1999 | 7.4 | 17,127 | 75,000 |

Source: Bağci *et al.* (2000).

Turkey is located on the active Alpine–Himalayan fault (Atac 1995) – see Figure 13.1 for an earthquake zoning map for Turkey. Some 55 earthquakes in this century have caused 70,000 deaths. The last ten years have seen three urban earthquakes: Erzincan, 6.8 on the Richter scale in 1992; Dinar; 6.0 in 1995; and Adana-Ceyhan, 6.3 in 1998. As can be seen from Table 13.2, the Marmara earthquake was the strongest of the last 30 years.

Immediate blame for Marmara was directed at construction contractors. However, all those who had a role in the building process played a part in this disaster. Vulnerability is linked to the economic, political and social structures in the country following a period of fast economic growth in the 1980s, which increased migration from rural to urban areas and put extra demands on housing provision. It is impossible, and unethical, to single out any one group as responsible. The disaster should be perceived as the manifestation of unresolved development challenges. This is imperative because of the close interaction between disasters and development, which should direct all strategies for dealing with them.

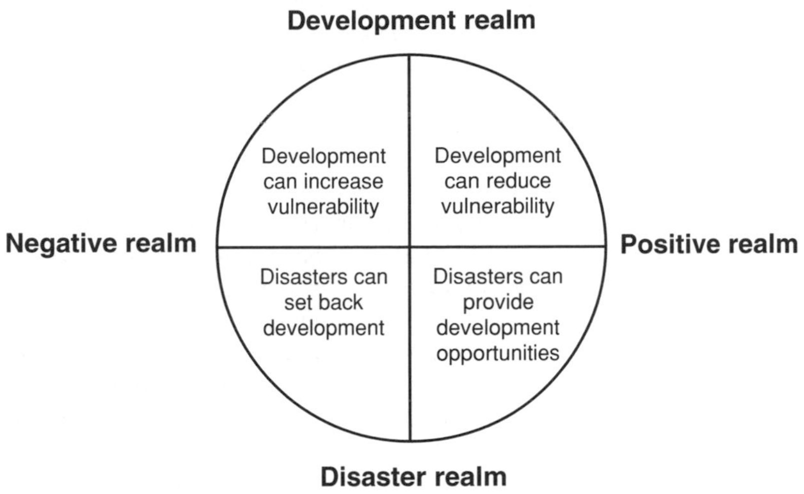


*Figure 13.2* The relationship between development and vulnerability to disasters.

Source: UNDP–DHA (1994: 8).

## Disasters and development

The interaction between vulnerability to disasters and socio-economic development is not well understood. There are negative and positive aspects to this relationship, subdivided below into four relationships between disaster and development (Sirleaf 1993; UNDP–DHA 1994) (see also Figure 13.2).

### *The positive realm*

1 Sustainable development can reduce vulnerability by addressing the root causes of disasters and the lack of access to economic and political tools. 'Mitigation is most effective as part of a medium- to long-term development program which incorporates hazard reduction measures into regular investment projects' (UNDP–DHA 1994: 8–9). Mitigation strategies can be classified as structural, including the construction of hazard resistant buildings, windbreaks and dams; or non-structural, including building codes, land-use policies and procedures for forecasting and warning. Ironically, post-disaster environments are often seen to be too turbulent to implement developmental programmes.

2 Disasters can provide new opportunities for sustainable development. This requires a development approach that is both sensitive to opportunities and able to respond by designing reconstruction programmes to incorporate such opportunities (Barakat 1993; Sirleaf 1993). Supporting this view, UNDRO (1992: 19) states that: '[d]isasters often create a

political and economic atmosphere wherein extensive changes can be made more rapidly than under normal circumstances . . . The collective will to take action is an advantage that should not be wasted.'

### *The negative realm*

1 Disasters can set back development initiatives by the loss of resources, interruption of programmes, impact on investment climate, impact on the non-formal sector, and political destabilization (UNDRO 1992). Boutros-Ghali (1995: 34) articulates this, asserting that '[b]ecause natural disasters can quickly devour hard-won achievements, planning must focus on ways to cushion the inevitable shocks, so that social structures will not be irreparably damaged, economic initiatives will not be forever set back, and natural disaster victims will not be condemned to perpetual dependence on external assistance'.
2 Poorly planned development programmes increase vulnerability. If development efforts are not appropriate to existing environmental factors, and their impacts on the environment have not been assessed properly, they can increase vulnerability to disasters.

These four areas of the disaster/development relationship are represented in Figure 13.2.

Quarantelli (1978), Davis (1978, 1986), Anderson and Woodrow (1989), and Blaikie *et al.* (1994) all present disaster as occurring when its two main components, hazard and vulnerability, coincide in time and place. Developing this argument, Blaikie *et al.* (1994) argued that distant root causes such as limited access to power, structures and resources, and ideologies in terms of political and economic systems, lead to dynamic pressures, such as rapid population growth and urbanization, foreign debt, war, lack of ethical standards in public life and environmental degradation. A consequence of these pressures is that populations are exposed to unsafe conditions where a fragile physical environment and local economy unite to produce vulnerability.

In this tradition, Barakat and Davis (1998) recommended confronting urban vulnerability through a 'risk spreading' method where a number of mitigation strategies are adopted in parallel. The main elements, with measures to be taken, are as follows. First, legislation to establishment of national, provincial and local preparedness plans. Second, education and training, including public awareness programmes, the inclusion of relevant safety elements in the curriculum of key professions and regular training programmes for decision-makers at governmental and non-governmental levels. Third, promoting insurance as a mechanism for inspecting and approving constructions as well as covering against loss. Fourth, municipalities should set examples of good practice in the construction of public buildings and infrastructure, providing a wide range of key lifeline buildings such as hospitals, airports, ports, schools and police stations, and infrastructural lifelines of water, electricity, communications and roads into the urban fabric. Fifth, a six-stage sequence of actions for disaster planning is proposed: (1) inception of disaster management, (2) risk assessment, (3) defining levels of acceptable risk, (4) preparedness and mitigation planning, (5) testing the plan, and (6) feedback from lessons learned using environmental impact analysis, hazard impact analysis, off-site safety plans and on-site safety plans.

Taking these strategies as a model, the geography of development in Turkey and the causes and consequences of the Marmara earthquake are explored below.

## Development in Turkey

Turkey's Human Development Index ranking is 86 out of 174 countries, falling in the medium human development group (UNDP 1999). Some 14 per cent of the population live in poverty (UNDP 2000). However, its US$6,400 per capita income puts the Turkish economy in sixth place in Europe. Population growth is 1.8 per cent, half the figure of eight years ago (Cem 2000). This section highlights the most important development challenges since the Second World War: economic liberalization and over-urbanization, plus political instability, corruption and a lack of accountability.

### *Economic liberation and over-urbanization*

Six phases of change in Turkey's attitudes to its national space from 1923 to 1980 were identified by Beeley (1998). First, 1923–1930: the young Republic's territory as undivided space emphasizing national integration centred on major urban areas within a hierarchy focused on the new capital, Ankara. Second, 1930–1939: the Depression era was characterized by the Turkish state managing change by centralized industrial plans and development initiatives. From the 1930s until 1980, import-substitution industrialization by means of public enterprises and development planning continued. Third, 1939–1950: military and economic links with the West were established through NATO and the US's Marshall Plan. Fourth, 1950–1960: a multi-party system meant a change from nationally focused priorities such as emphasis on 'regional development' to spatially concentrated 'point' impact. However, the overall acceptance of the notion of public sector-led economic development remained.

Fifth, 1960–1980: the military take-over in 1960 brought an era of co-ordinated planning through a series of five-year plans. These stressed the need for public–private mix. Contrasting levels of development, urban and rural, and particularly east and west, were areas for particular concern. Unequal development caused migration both abroad and into the cities of

Turkey. Sixth, after the military take-over on 12 September 1980, Turkey entered a new wave of liberal trade and payment regimes under the premiership of the late Turgut Özal. National economic growth was obtained but at the cost 'of accumulating a large external debt which climbed more than fivefold from less than $10 billion in 1980 to more than $50 billion in 1990' (Boratav 1990: 223). Income distribution worsened as 'real wages declined by as much as 34 percent and the intersectoral terms of trade turned against agriculture by more than 40 percent from 1977 until 1987' (Boratav 1990: 223). Labour repression by the military and the generous flow of foreign capital supported Özal's policies, but old-style populist policies after 1987 caused payment deficits to worsen, increasing the rate of inflation to its pre-1980 levels of over 70 per cent per annum (Owen and Pamuk 1998: 117–122), disrupting macroeconomic balances, eroding social structures, worsening income distribution and indirectly creating a fertile ground for corruption (Baran 2000).

Against the background of a liberalizing economy from the 1980s, the rate of migration from rural areas to towns and cities has accelerated. The 1932 plan for Ankara envisaged a population of 300,000 within three decades, today it is over 4 million (Beeley 1998). In 1999, the urban population in the country reached 70 per cent. A direct consequence of this over-urbanization has been the failure of housing provision mechanisms in Turkey to cope with increasing demand. This has led to the construction of informal settlements and a growth in informal economic activities in big cities such as Istambul, Ankara and Izmir. Informal settlements often avoid demolition by constructing a mosque and can gain basic infrastructure through the de facto recognition that comes from acting as a vote bank for politicians. Inhabitants expect they will gain planning permissions before each general election. Unplanned growth has also generated land speculation and led to the development of public land.

### *Political instability, corruption and lack of accountability*

Since 1984, the war between the PKK (Partiya Karkeren Kurdistan) and the Turkish security forces has displaced thousands of people from their homes and livelihoods and caused around 30,000 deaths at an estimated cost of US$7 billion a year to the state. The Turkish political life has always been very turbulent, witnessing three military take-overs; ideological, sectarian and ethnic polarization; human rights violations; threats posed by fundamentalist Islam against the state's secular system; links between state, right-wing terrorists and organized crime; and assassinations on secularist and leftist journalists, academics, politicians and trade unionists. Corruption at all levels of society has found an ideal breeding ground worsening during the liberalization process because of the lack of a strong and independent judiciary, political accountability and financial transparency (Baran 2000).

## Lessons for disaster management practice in Turkey

### *Legislation and training*

The 1997 Turkish Earthquake Resistant Design Code for Buildings, which is an adaptation of the Uniform Building Code in California, is sophisticated and strict, and multi-storey buildings in Turkey should all be earthquake resistant. However, recent experience shows that they are not. In the Dinar earthquake one-third of all dwellings were damaged at medium and heavy levels. According to the Earthquake Engineering Research Institute (EERI 1998), this was partly because of unsupervised construction, with work rarely being inspected. Another reason was that construction had taken place on unsuitable, soft ground known on the basis of historical earthquake records and geological surveys as having a high potential for earthquakes. Obviously such consequences were not scientifically acknowledged and their incorporation into the development of housing and industrial areas was 'neglected' by national and local planning authorities.

Assessment by the Istanbul Technical University (ITU 1999) on civil engineering aspects of the Marmara earthquake emphasized that disregard of Turkey's strict building and earthquake safety regulations was the main reason why engineered buildings were affected so badly. EQE International supported these conclusions after carrying out a technical survey in the affected area, pointing out that

> [m]ost of the buildings did not meet the design requirements of the code and included details that are not earthquake resistant . . . Many of the buildings were built with poor and inappropriate construction materials and utilized poor workmanship. Many buildings were knowingly allowed to be built on active faults and in areas of high liquefaction potential. Many buildings were not engineered, but built according to past experience.
>
> (EQE International 1999)

The relationship between contractors and engineers is important and two-dimensional. Middle East Technology University's Disaster Management Implementation and Research Center (1999) notes that high competition for contract design work and low levels of remuneration have reduced engineers' willingness to develop professional competence. EQE International (1999) note that design engineers, who are typically employees of the contractor, tend not to inspect on-site construction; this failure in oversight allows on-the-spot modifications, which can compromise the earthquake resistance of the building.

A further critical aspect to this relationship is the inadequacy of the control mechanisms of local municipalities. In building code enforcement, Coburn and Spence (1992: 139) refer to the municipal engineer as 'the front-line soldier in the community's battle for earthquake protection', and if that is the case it seems that the battle in Turkey has so far been mostly lost. Some contractors there are corrupt so it is essential that local municipalities have the financial resources and trained personnel to be able to inspect their work. However, Turkey is not a high-income country and municipal councils usually do not have the resources to employ adequate numbers of civil engineers as inspectors. Even if they do, those civil engineers are often poorly paid so it is difficult for local municipalities to attract experienced engineers who can achieve more than a simple checking of the basic calculations of building projects. In addition, it is difficult to see how to achieve proper building inspection, given the possibility of corruption in obtaining building permission through bribes and political favours.

### *Increased public awareness*

After each previous earthquake politicians asked the public for solidarity, patience and compassion. Each time, the Turkish public responded positively. But this time the establishment in Ankara got a different reaction, with the public demanding to know why the 'act of God' became a disaster for them. After the earthquake in Adana some people went to court to sue their building contractors, but these demands for justice were probably not strong and visible enough to affect the practice of 'business as usual' with building safety in Turkey. Although some initiatives by civil society, academia and the Turkish media tried to focus on the causes and consequences of the Marmara earthquake they will need to be persistent since public concern in Turkey tends to move on quickly. The close proximity of the earthquake area to the media headquarters in Istanbul was also important. Due to public pressure the establishment now take the matter of earthquake safety and disaster management more seriously and incorporate lessons learned from the long list of Turkish earthquakes into preparedness and mitigation. But why now? There are two reasons. First, the unprecedented scale of damage inflicted by the Marmara earthquake. Second, this was a disaster whose victims were mainly urban, making it difficult for politicians, local municipalities, building contractors and civil engineers not to accept responsibility.

The level of death and destruction in Marmara has created a new level of public awareness in Turkey, with demands for the enforcement of regulations concerning construction practices to minimize risk, quality control during new constructions and the retrofitting of buildings and houses at risk. More importantly increased earthquake consciousness seems to have shaken the state's institutional structures and its complacency. Özyaprak (1999) argues that the earthquake was positive in that society in Turkey has 'discovered' its civil identity and realized the importance of civil initiative as a powerful tool in making social and economic changes take place. Supporting this view, Incioglu (1999) claims that it has initiated the process of questioning the overall viability, effectiveness and organizational structure of the state, particularly the role of the army. The image of the state as a protective father among the population in Turkey has collapsed totally. The public is fully aware of the overall political system and economic policies which led to this disaster, and are demanding a reformed state system which has high organizational capacities, is effective and democratic.

### *Insurance*

The recent initiation of a general insurance scheme for buildings is expected to play a significant role in ensuring that they are built according to building safety regulations. It is likely that insurance companies would refuse to provide insurance for a building that is not earthquake-proof, or would at least ask for high premiums. Experience in Turkey shows that unless there is a financial incentive, regulations and rules are very difficult to implement. The government has already made the necessary legal arrangements for the enforcement of building insurance for all new constructions. The establishment of chartered construction inspection firms is seen as a necessity, a radical shift moving the responsibility of inspection from local authorities to chartered firms to avoid the possibility of corruption between construction contractors and local authorities. However, the oath will not be made by these firms, but by such professionals as architects and engineers working there. Considering that these measures have been taken against the corruption experienced in construction processes, which are carried out by professionals but not firms, these legal changes need to be part of an overall planning discipline. The inspection process needs to be independent and led by science not faith. However, this initiative should be considered as a major step in the right direction, requiring further consolidation with other tools of urban planning.

Bil (1999) reports that insurance companies are not willing to accept full responsibility for ensuring the earthquake safety of buildings. It is crucial that a system of independent and chartered monitoring and control mechanisms, from the planning stage to obtaining building permission, should be instigated, and this checking mechanism should continue throughout the construction. Local authorities should be equipped with the power to stop constructions if they are found to be defective. Overall, the encouragement of building insurance seems to be an effective, inexpensive and sustainable tool for ensuring building safety. For example, the cost of ensuring an apartment flat with a value of TL 10 billion (US$18,000) is estimated to be around TL 40 million (US$70) a year. However, this should be done in parallel with the establishment of necessary monitoring and control firms as otherwise the mechanism would not prove to be efficient and effective.

The unpredictable frequency and devastating impacts of earthquake disasters also discourage the international insurance community from providing earthquake insurance in seismically active areas. Heywood (1995) presents the relationship between risk and insurance in Table 13.3, which

*Table 13.3* The relationship between risk and insurance

| *Severity* | *Frequency* | |
|---|---|---|
| | *Low* | *High* |
| High | Classic higher-risk insurance | Tendency to become uninsurable |
| Low | Deductible or self-insurance | Risk management or pay-back insurance |

Source: Heywood (1995).

illustrates the choice available. This shows that low frequency and low severity are manageable but that high frequency and high severity will tend towards the uninsurable, indicating the challenges faced by insurance companies. Based on historical records of earthquakes in Istanbul it is estimated that the city will be struck by a major earthquake in the near future. The physical and economic losses caused by recent earthquakes in Los Angeles (USA) and Hanshin (Japan), which are comparable to Istanbul in terms of their population, economic concentration and building density, were as high as US$200 billion (Erdik 1995). Bearing in mind that buildings affected by recent earthquakes in Turkey showed low earthquake-resistant performance, the cost of covering such losses could easily overwhelm an insurance company.

### *Urban planning and management*

Setting examples of earthquake-proof constructions by governmental and local authorities was suggested by Barakat and Davis (1997) as an effective urban management tool. Experience in Turkey shows that the state is far from setting any kind of example in the construction of public buildings and infrastructure. It was the 1992 earthquake in Erzincan that drew public attention for the first time to the fact of low earthquake safety of public buildings. According to Isikara (1999), in addition to 35 destroyed public buildings, several schools, a major hospital, residential buildings for judges and attorneys, and two big hotels collapsed in Erzincan. The situation with public buildings in the Dinar earthquake was even worse as 25 out of 55 collapsed, raising further questions with regard to the corruption in the civil service. At the Marmara earthquake, similar experiences with public buildings were repeated, as 43 schools in the earthquake-affected area collapsed while 377 of them were damaged (DPT 1999).

The Marmara earthquake inflicted tremendous damage on the industrial production capacity of the area, with large losses of buildings, machinery and equipment, stock and qualified personnel. According to the DPT (1999) the loss of industry in the area in monetary terms is estimated at US$600–700 million, which equals a decrease of 1.6 per cent in the country's annual growth, and the reconstruction of damage inflicted on public heavy industry was around US$220 million. The production and market losses from these industries were estimated at around US$630 million. The TUPRAŞ Refinery fire, which burned out of control for several days, required the evacuation of everybody within a three-mile radius as the area around it was under imminent danger from possible explosions. The situation was particularly dangerous as the refinery was very close to other heavy industries such as petrol stations and fertilizer production plants. The over-concentration of public and private heavy industries such as petrochemical plants, paper mills, car manufacturers, pharmaceutical firms and cement plants in an earthquake-prone area and near densely populated urban areas underlines an important reality in Turkey, which is the phenomenon of unplanned regional and urban development.

According to the Middle East Technical University's Disaster Management Centre (1999), urban development law is narrow in its remit to address the building stage only. It is further explained that 'Disaster Law is obsolete in many ways, and contributes to the myth of the omnipotent state that will intervene in the event of any disaster, rebuild a dwelling for every citizen who loses one, plus a workshop for heads of households if they held a deed to one before'. It was claimed that the explanation for this type of complacency can be found in the overall psycho-social characteristics of the Turkish society: 'It is not surprising that in a culturally fatalistic society, this makes consumers blasé with regard to the structural quality of buildings in which they entrust their own lives and their families to divine intervention.'

Balamir (1999) implies the concept of a culturally fatalistic society in his disaster management framework, in which fatalistic and self-esteemed approaches form the two opposing sides of the spectrum. He points out that the main reason why Turkey has experienced shortcomings with its disaster management strategies is a result of this overall fatalistic approach. He claims that fatalism is compatible with a policy preference for relief over mitigation and preparedness.

### *Disaster response strategies*

The aftermath of the Marmara earthquake highlighted the extremely limited capabilities of the General Directorate of Civil Defence of Turkey to provide a quick and effective response to the needs of disaster-stricken people. In a country like Turkey, where earthquakes are a part of life, it is beyond belief that the state cannot organize an effective civil defence system. In contrast to the state institutions' poor response to this earthquake, the involvement of civil society organizations in the provision of emergency aid and services was particularly effective. There was major participation by the different non-governmental organizations in rescue works of those trapped under collapsed buildings, and the public in Turkey has started to question the state and its institutions in their response to the needs of earthquake-affected people whilst the civil society organizations' popularity with the public reached its peak. Kemal Demir, Director of the Turkish Red Crescent, which was criticized for being incompetent in its response to the disaster, had to bow to public pressure and resign from his post on 7 October 1999. The government in Ankara should immediately start to think of ways of improving the Civil Defence organization's structure and capacities. The Directorate currently has main units in Ankara, Istanbul and Erzurum, and the total number of its personnel, both administrative and technical, is only one hundred. Despite this, the Directorate claims that Urgent Rescue and Relief Teams were formed in each province, of which there are 80, consisting of 50 to 150 personnel, though the Marmara experience showed that these were only visible on paper without training and equipment. It is in this context that the international community could have a significant role. The transfer of knowledge in rapid rescue methods and the designing of effective disaster preparedness plans are areas where the international community can provide Turkey with the necessary resources for capacity strengthening.

## Conclusion

The causes and consequences of the Marmara earthquake have been evaluated in this chapter. This has highlighted a series of challenges in terms of responses to disaster management needs, but also the existence of a number of opportunities which have emerged or been exposed by the disaster itself. Considering that earthquakes are part of everyday life in Turkey, it is imperative that these findings are incorporated into the country's disaster management strategies by utilizing the special characteristics of the post-disaster social, institutional and political environment. The following are the salient points for consideration:

1 Sophisticated and strict regulations for earthquake-resistant design on their own do not ensure earthquake-proof buildings.
2 Earthquake proofing cannot be ensured by the implementation of civil engineering measures only; soil conditions must also be considered.
3 A wide spectrum of civil engineering failures means that a large proportion of the existing housing stock in Turkey is likely not to be resistant to earthquakes.
4 There is an urgent need for the civil engineering sector to review its role in the handling of this disaster. The convening of a Forum for Civil Engineering professionals could address the challenges faced by bringing together academics, researchers, practitioners and representatives of local authorities and should be considered urgently.
5 The entire construction process needs to be restructured in an integrated and holistic manner. The availability of highly qualified professionals and distinguished academics forms a large pool of talent which has yet to be tapped by a government-led initiative.

6 As a result of increased public awareness, coupled with media focus on the issue, government is likely to take earthquake disaster management more seriously, at least in the short term. Academic and other civil society pressure groups should keep up the pressure.
7 The earthquake served as a trigger for the public in Turkey to rediscover the scope and power of its civil society.
8 Demand from potential home buyers for the provision of earthquake-resistant buildings should be used to create a popular lobby for reform.
9 A comprehensive building insurance scheme, incorporating the establishment of chartered construction inspection firms, is needed. The insurance sector should be in a position to take up this role as long as the system is supported with other necessary institutional structures. The main hurdle is the need to generate sufficient political will to carry out the required legal and regulative reforms.
10 Government and local authorities are far from setting examples in the construction of earthquake-resistant buildings. Lack of control mechanisms, corruption, political favouritism and the complacency of state institutions are a few of the complex root causes behind this fact.
11 Urban planning law, disaster management law and regulations on building construction need to be restructured in order to make them more interconnected and co-ordinated in their overall roles in disaster preparedness and management.
12 The reality of living between two earthquakes should be understood by all layers of the society. Institutional frameworks responding to earthquake disasters should be restructured and equipped with adequate financial and institutional resources.
13 The utilization of the army as a pool of physical and human resources in responding to disasters should be reconsidered in order to increase their contributions towards the overall aims of public awareness and emergency response.
14 The realization of financial, institutional and legal assistance to civil society organizations working in disaster response is imperative in order to encourage the creation of a wider disaster-response network.

## Acknowledgement

This chapter draws on the author's joint article with Dr Sultan Barakat entitled 'After the Marmara Earthquake: Lessons for Avoiding Short Cuts to Disasters', published in *Third World Quarterly* 21 (3), 2000.

## References

Anderson, M. and Woodrow, P. (1989) *Rising from Ashes: Development Strategies in Times of Disaste*r, Paris: UNESCO.

Atac, S. (1995) 'Quake begins to shake official complacency', *Turkish Daily News*, October 7, p. B1.

Bağci, G., Yatman, A., Özdemir, S. and Altin, N. (2000) 'Türkiye'de hasar yapan depremler (Damaging earthquakes in Turkey)', *Deprem Araştirma Bülteni* 69, 113–126.

Balamir, M. (1999) 'Afet zararlarinin azaltilmasi amaciyla planlama ve yapilanma sureclerinin yeniden orgutlenmesi ve ODTU onerisi (Planning to reduce disaster damages and reorganization of construction processes, and ODTU proposal)', *Mimarlik* 288, August, 2–16.

Barakat, S. (1993) 'Reviving war-damaged settlements: towards an international charter for reconstruction after war', Unpublished D.Phil. thesis, IoAAS, University of York.

Barakat, S. and Davis, I. (1998) 'Disaster preparedness for Palestine', in A.B. Zahlan (ed.) *The Reconstruction of Palestine: Issues, Options, Policies and Strategies*, London: Kegan Paul.

Baran, Z. (2000) 'Corruption: the Turkish challenge', *Journal of International Affairs* 54 (1), 127–146.

Beeley, B. (1998) 'On the geography of development in Turkey', in E. Watkins (ed.) *The Middle Eastern Environment*, London: St Malo Press.

Bil, F.C. (1999) '40 milyon liraya yapi sigortasi (Building insurance for TL 40 million)', *Sabah* supplement: 'Insaatta Yeni Donem (New Era in Construction)', 28 September, 11.

Blaikie, P., Cannon, T., Davis, I. and Wisner, B. (1994) *At Risk: Natural Hazards, People's Vulnerability and Disasters*, London: Routledge.

Boratav, K. (1990) 'Inter-class and intra-class relations of distribution under "structural adjustment": Turkey during the 1980s', in T. Aricanli and D. Rodrik (eds) *The Political Economy of Turkey: Debt, Adjustment and Sustainability*, London: Macmillan.

Boutros-Ghali, B. (1995) *An Agenda for Development*, New York: UN.

Cem, I. (2000) 'How Turkey can contribute to Europe's brighter future', in The Philip Morris Institute for the Public Policy Research, *Has the EU Enlargement Process Lost Its Way?*, Brussels: The Philip Morris Institute for the Public Policy Research.

Coburn, A. and Spence, R. (1992) *Earthquake Protection*, Chichester: John Wiley and Sons.

Davis, I. (1978) *Shelter After Disaster*, Oxford: Oxford Polytechnic Press.

Davis, I. (1986) 'Lessons from reconstruction after natural disasters for cities recovering from bombing and civil strife', *Proceedings of an International Conference on Reconstruction of the War-damaged Areas*, 6–16 March, University of Tehran (I.R. Iran).

Devlet Planlama Teskilati (DPT) (1999) 'Depremin ekonomik and sosyal etkileri: muhtemel finans ihtiyaci (Economic and social impacts of the earthquake: possible financial needs)', Available online: <http://ekutup.dpt.gov.tr/deprem> (accessed 19 October 1999).

Earthquake Engineering Research Institute (EERI) (1998) 'Dinar aftershock tests retrofitted buildings, EERI Special Earthquake Report, September 1998', Available online: <hhtp://www.eeri.org/Reconn/Dinar98/Dinar98.html> (accessed in October 1999).

EQE International (1999) 'Izmit, Turkey earthquake of August 17, 1999 (M7.4): An EQE briefing', Available online: <http://www.eqe.com/revamp/izmitreport/index.html> (accessed in October 1999).

Erdik, M. (1995) 'Istanbul icin bir deprem senaryosu (An earthquake scenario for Istanbul)', *Cumhuriyet*, 28 August.

Heywood, J.R. (1995) 'Natural hazards as problems for insurers', Paper presented at the UK International Decade for Natural Disaster Reduction (IDNDR) Seminar held on 6 October 1995, London.

Incioglu, N. (1999) '"Devlet baba" imaji coktu (The image of "father" state collapse)', *Milliyet*, 31 August, 18.

Isikara, A.M. (1999) 'Depremle yasamak gercegi (The reality of living with earthquakes)', *Sabah*, 30 September.

Istanbul Teknik Universitesi (ITU) (1999) On Degerlendirme Raporu (Pre-Assessment Report), Unpublished report, 24 August.

Middle East Technical University (METU) (1999) 'A simplified analysis of the earthquake vulnerability of the housing stock in Turkey', Available online: <http://www.metu.edu.tr/home/wwwdmc/simp.html> (accessed 10 December 2000).

Owen, R. and Pamuk, S. (1998) *A History of Middle East Economics in the Twentieth Century*, London: I.B. Tauris.

Özmen, B., Nurlu, M. and Güler, H. (1997) 'Cografi bilgi sistemleri ile deprem bolgelerinin incelenmesi (Earthquake Zone Investigation with the aid of GIS)', Available online: <http://www.dask.gov.tr/dask/deprem/harita.html> (accessed December 2001).

Özyaprak, S. (1999) 'Yapisal degisim zamani mi? (Is it time for structural changes?)', *Financial Forum*, 11 August, p. 9.

Quarantelli, E.L. (1978) *Disasters: Theory and Research*, Oxford: Sage.

Sirleaf, E.J. (1993) 'From disaster to development', in K. Cahill (ed.) *A Framework for Survival: Health, Human Rights and Humanitarian Assistance in Conflicts and Disasters*, New York: Basic Books and The Council of Foreign Relations.

United Nations Development Programme (UNDP) (1999) *Human Development Report 1999*, Oxford: Oxford University Press.

United Nations Development Programme (UNDP) (2000) *Human Development Report 2000*, Oxford: Oxford University Press.

United Nations Development Programme – Department of Humanitarian Affairs (UNDP–DHA) (1994) *Disasters and Development* (Disaster Management Training Programme, 2nd edition), Prepared by R.S. Stephenson, New York: UNDP.

United Nations Disaster Relief Organization (UNDRO) (1992) *An Overview of Disaster Management* (Disaster Management Training Programme, 2nd edition) Geneva: UNDRO.

# 14 Ecological reconstruction of the upper reaches of the Yangtze river

*Chen Guojie*

## Introduction

The Yangtze is the longest river in China, and the third in the world. It originates from the north of the Danggula Mountains in Qinghai Province, dissects China's interior and then flows into the Pacific Ocean at Shanghai; it is 6,300 km long. The area of the Yangtze valley is 1,800,000 square kilometres, accounting for about 18.8 per cent of the land area in China. Some 34 per cent of the population of China live in the Yangtze valley. The output value of agriculture and industry accounts for 32 per cent and 28 per cent respectively of that of China, which makes the Yangtze valley one of the most developed belts in the country. However, due to the deterioration of the ecological and environmental functions of the valley, natural disasters such as floods have become more frequent in the last 50 years. This threatens the ecological safety of the Yangtze valley.

The upper reaches of the Yangtze river are sensitive to environmental change, and disturbances here can result in disasters along the valley. For this reason it is the attempts to restore the ecological health of the upper reaches of the valley that this chapter discusses. The upper reaches begin at Yichang, Hubei Province (see Figure 14.1). They cover 1,054,000 square kilometres, some 58.9 per cent of the Yangtze valley. The population is 163 million, accounting for 40 per cent of that of the valley. The largest settlements are provincial administrative units at Qinghai, Xizang, Sichuan, Ganshu, Shanxi, Yunnan, Guizhou, Chongqing and Hubei. Ethnic minorities such as the Tibetan, Qiang and Yis are the majority population (Yang Dingguo 1996).

This zone is biologically diverse, supporting more than 4,100 species of medicinal plants. The giant panda, red panda and snub-nosed monkey are some of its most famous rare animals. The ecological and cultural importance of the region is reflected in its many reserve areas. The Hengduan (Transverse) Mountain Region has been defined one of the 25 protected hot spots of biological diversity by Conservation International. The Ermei Mountain–Leshan figure of Buddha, Jiuzhaigou (a beauty spot), Huanglong Temple (an outstandingly beautiful calcified landscape), Du Jiang weirs (given

*Figure 14.1* The upper reaches of the Yangtze river.

world-class cultural heritage status) and Lijiang (an ancient town) are of especial interest. More than this, there are abundant natural resources in the area (see Table 14.1).

The river runoff of the upper reaches accounts for about 50 per cent of total runoff in the Yangtze Valley, with rainfall from May–August making up more than 70 per cent of total rainfall for the year. The upper reaches, with high mountains and deep valleys, run from Qinghai–Tibet, with an

*Table 14.1* Natural resources, upper Yangtze river

| *Resource* | *Proportion of known Chinese reserves (%)* |
|---|---|
| River runoff | 17 |
| Theoretical reserves of waterpower | 33 |
| Exploitable minerals: | |
| alum, titanium | 46 |
| strontium | 66 |
| mercury, glare | 70–90 |
| natural gas | 60 |
| phosphoric | 40 |
| aluminium | 28 |
| halite and pyrites | 25 |
| Forest land | 18.8 |
| Timber reserves | 6 |
| Grass land | 20 |

altitude of over 4,000 kilometres, into the Sichuan basin at 20 metres altitude. The gradient ratio is so high that severe flooding easily occurs in the middle reaches of the Jianghan plain. Flood control, water supply and water quality depend much on the change of water level in the upper reaches and the capacity of controlling and regulating water, which has much to do with water conservancy construction and forest vegetation recovery in the upper reaches. Calculations show that if the forest acreage were to recover to 30–40 per cent, meaning an additional 100,000–200,000 square kilometres of forest being cultivated, then a reservoir with a volume of 15–20 thousand million cubic metres could be built.

The upper reaches region spans three climate zones: altiplane, north subtropical and middle subtropical. Climate changes here are greatly affected by global circulation: including westerly circumfluence, south-westerly monsoon, south-easterly monsoon and monsoon from the Qinghai–Tibet plateau. The Qinghai–Tibet plateau, the start-up area of the monsoon from the Orient, is so sensitive to global climate fluctuations that regional weather patterns can be predicted from it. This makes the upper reaches a key area for the study of the impacts of global climate change. Observed environmental changes in the upper reaches have included a decrease in annual rainfall, the retreat of glaciers, a moving up of the snow line, and a drying up of grassland and of river valleys (Peng Puchu and Fu Chen 1999).

The population of national minorities is 14.6 million, accounting for 23 per cent of that of China. The region is below the mean level of economic development in China. It lags far behind coastal areas due to its poor communications, harsh conditions of production and the lack of quick access to information. Here live 34.5 per cent of those below the poverty line in China. As China opens up to Western-style development the need to keep pace with a rapidly expanding coastal economy is even more important to prevent growing economic inequality. Present economic development in the upper reaches remains based on the exploitation of natural resources, resulting in acute and complex relations between the population, resources and environment. The big challenge is to harmonize economic development with ecological construction. Priority concerns include degradation in the dry river valleys of the Jinsha, Yalong, Dadu and Min rivers, desertification of grassland in areas of east Qinghai–Tibet plateau and, in the Yangtze river source region, soil erosion in basin and hill areas, lithification of karst areas, resettlement of people made homeless by the Three Gorges Reservoir project, and the protection of natural forest in areas with high mountains and deep valleys, to name just a few (Chen Guojie 2000)!

## Ecological degradation

A flood causing direct losses of ¥300,000 m (RMB) (US$36,000 m) occurred in the Yangtze valley from June–August 1998. With eight peaks it almost completely destroyed the embankments of the middle reaches of the Yangze river. This natural disaster impacted on almost all the valley from Sichuan, Chongqing municipality to Shanghai municipality. This event should be seen as a caution to the seriousness of ecological and environmental degeneration in the Yangtze river's upper reaches. The underlying environmental changes that culminated in the flood are discussed below.

### *Loss of natural vegetation cover in the upper reaches*

Large-scale deforestation has occured since the middle or late 1980s, with forest acreage decreasing from 30 per cent to about 5 per cent of land area at its worst. Reforestation and conservation measures have increased the proportion of land under forest to 20 per cent, but the new forest stands are less biologically diverse and have a reduced ecological function (Bao Weikai and Chen Qingheng 1999).

Grassland concentrated in the Yangtze river source region and the northwest part of Sichuan has also suffered serious degradation. Degraded and desertified grassland occupies 18.9–27.7 per cent of the total grassland area in the Yangtze river source region and is linked to 71,000 square kilometres at risk of soil erosion. In the northwest of Sichuan, 40–60 per cent of the total grassland area has been degraded. It is thought that rapid growth in rodent populations is partly to blame for this.

### *Soil erosion and flooding*

Soil erosion in the upper reaches region affects 393,000 square kilometres – more than one-third of the total area. On average 1,568 million tonnes of soil are lost each year, equivalent to a 30 cm layer of soil over 387,000 ha. The storage capacity of water reservoirs in the Yangtze valley are decreasing by 1,200 million cubic metres per year because of the resulting sedimentation. Sedimentation has been linked to the flood peak at Yichang and Changsha in 1998, the highest recorded in history.

### *Aggravation of natural disasters*

Drought does most damage of all natural disasters in the upper reaches region. In Sichuan province alone, since the 1980s, the reduction of grain output due to drought has been 2,000 million kg/acre (Tang Bangxin 1995). Meanwhile, water-conserving and flow-regulating capacity is lost as a result of the deterioration of natural vegetation cover, which brings about more frequent flooding and greater damage. In Sichuan decadal flood frequency has increased since the 1950s, with the floods of 1981 and 1989 being the most serious to date (Tang Bangxin 1995). In the upper reaches alone, direct economic losses due to the 1998 flood amounted to US$4,693 m, more than twice that of 1981 (Chen Guojie 2000).

The year 1998 was a bad one for catastrophic landslides and debris flows. Catastrophic events killing or injuring more than 200 people, or producing direct economic losses in excess of US$1.2 m, occurred ten times in the upper reaches. Although the data is incomplete, over 1,000 large-scale landslides, 150 collapses and 3,400 debris flows were recorded in 1998. Real frequencies are likely to be much higher.

### *Deterioration of the water environment*

Though the upper reaches of the Yangtze river are rich in water sources, water scarcity and pollution are becoming more and more serious. On the one hand, river runoff is decreasing due to climate changes. Rates for the upper reaches of the Min river have decreased by 14 per cent in the last 50 years. Meanwhile, drought river valleys such as the Jinsha river, the Yaling river, the Dadu river and the Min river have a greater tendency to dry up. The contraction of wetland areas in the north-west part of Sichuan Province also points towards drying up and a heightened risk of desertification (Sun Guanyou 1998).

The annual amount of waste water flowing into the Yangtze river now reaches 42,600 m tonnes, not including agricultural runoff and acid rain. Water quality in principal tributaries, such as the Tuojiang river, and of sections of the mainstream, such as the Three Gorges Reservoir area, is below the Chinese government's standard of third-class water quality. Particularly high levels of pollution have been found around cities such as Chongqing, Yibin, Luzhou, Panzhihua and Wanxian. The drinking-water quality of towns is threatened by this water pollution. This contrasts with people living in mountainous areas who have difficulties acquiring sufficient quantities of drinking water and irrigation water all year round. Chengdu, the capital of Sichuan Province, which used to be rich in water resources, is now a water-scarce city (Fan Yiping and Chen Guojie 2001).

### *The causes of ecological degradation*

Ecological/environmental deterioration is caused by the interaction of nature, economy and society. The backward economy and poverty all over the region are both the cause and the result of ecological and environmental deterioration (Yang Dingguo 1996). The example of West Sichuan illustrates this in detail.

West Sichuan is a typical underdeveloped interior region of China. It includes three autonomous prefectures (Ganzi, Aba and Liangshan), two cities (Panzhihua and Leshan) and Ya'an prefecture. Land area is 331,879 km with a population of 15,200,000 in 1998 (*Statistical Yearbook of Sichuan Province* 1999). Forest lumbering has been banned since 1 September 1998. The region is located in the east of the Qinghai–Tibet plateau where the Jinshajiang, Yalongjiang, Dadu and Minjiang rivers converge. Nationally, it is one of the most important water source regions, and includes important forest and pastoral areas. Environmental changes here impact locally but also throughout the Yangtze valley.

In West Sichuan, contemporary problems involve water and soil erosion, grassland degradation, forest coverage reduction, river withering, drought attack, bio-diversity decrease and threats from mountain hazards (Table 14.2).

Poverty is very extensive in West Sichuan, with dependence on the environment linked to ecological deterioration. Eventually a vicious cycle of poverty and ecological loss occurs (Figure 14.2). This cycle is directed by the interaction of two secondary cycles: one socio-economic and the other ecological.

As is shown in Table 14.3, agricultural livelihoods dominate in West Sichuan; if cities are excluded the proportion of citizens engaged in agriculture

*Table 14.2* Deteriorating ecosystems in West Sichuan, 1999

| *Prefecture* | *Damaged grassland (ha)* | *Proportion of damage to the whole (%)* | *Water and soil loss (km²)* | *Forest cover, 1999 (%)* | *Forest cover, 1950s (%)* |
|---|---|---|---|---|---|
| Yaan | – | – | 4,616.05 | 43.4 | >50 |
| Ganzi | 204 | 23.3 | 54,500 | 11.3 | 12.8 |
| Arba | 100 | 50 | 29,830.89 | 21.3 | 34.4 |
| Liangshan | 54.84 | 23.4 | 29,500 | 28.6 | 60.0 |
| Leshan | – | – | 6,217.14 | 34.8 | >50 |
| Panzihua | – | – | 3,650.72 | 42.9 | 57.98 |

Source: Environmental planning (2000 to 2030) prefecture and city.

*Table 14.3* Social and economic development in West Sichuan, 1999

| *Prefecture* | *Area ($km^2$)* | *Pop. (10,000)* | *GDP million yuan* | *Farmers (10,000)* | *Per capita, farmer gross income (yuan)* | *Per capita, GDP* | *Farmer, total (%)* | *Farming labour, total (%)* |
|---|---|---|---|---|---|---|---|---|
| Ya'an | 15,314.0 | 148.91 | 67.0816 | 121.82 | 1831 | 4,503 | 81.76 | 83.82 |
| Ganzi | 152,629.0 | 87.629 | 23.5932 | 74.23 | 721 | 2,692 | 84.71 | 94.59 |
| Arba | 83,426.2 | 81.6465 | 33.4749 | 66.34 | 1227 | 4,108 | 81.25 | 88.21 |
| Liangshan | 60,115.0 | 393.9342 | 134.4087 | 347.38 | 1243 | 3,412 | 88.18 | 93.46 |
| Leshan | 12,826.0 | 345.1246 | 135.0210 | 265.63 | 2077 | 3,912 | 76.97 | 76.00 |
| Panzhihua | 7,434.4 | 101.80 | 108.3057 | 47.58 | 2381 | 10,639 | 46.74 | 88.52 |
| *W. Sichuan* | *331,744.6* | *1,159.135* | *501.8851* | *922.98* | – | *4,330* | *79.63* | *85.61* |
| Sichuan | 485,000.0 | 8,348.4 | 3,711.61 | 6,940.9 | 1843 | 4,356 | 83.14 | 72.01 |
| China | 9,600,000.0 | 1,196.5 | 82,054.0 | 92,216.3 | 2210 | 6,545 | 76.09 | 70.18 |
| Guangdong | 177,900.0 | 7,298.88 | 7,919.1 | 5,026.42 | 3629 | 11,728 | 68.86 | 57.45 |

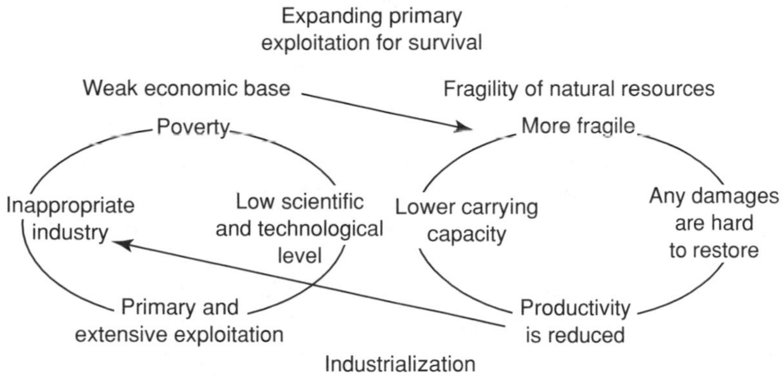


*Figure 14.2* The vicious cycle of poverty and environmental degradation, West Sichuan.

reaches 82.79 per cent. However, because farming practices remain largely traditional, farming offers only a very low per-unit yield. Therefore, within this agricultural region the amount of grain accessed by each household is lower than the state average. Together with poverty, high population and high extraction rates threaten the carrying capacity of agricultural lands and natural grasslands alike (Chen Guojie 1999a).

West Sichuan is a high altitude region (2,000–4,000 m above sea level) with a severely cold climate. Natural hazards are serious and frequent. Opportunities for developing agriculture are limited, with natural productivity being very low. Indigenous ethnic groups still apply predominantly traditional farming methods. Advanced science and technology have made little impact locally to accelerate social and economic development. For a long time, state development policy for West Sichuan has focused on forestry, mining and water management without much concern for environmental impacts. At the same time there was little investment in secondary and tertiary industry. Thus regional economic development policy has conspired to limit livelihood opportunities outside of agriculture and other primary industrial activities, providing only relatively low incomes. Add to this a population growth rate which for several years has been increasing above the rate of increase in agricultural productivity and it is not difficult to imagine the situation of population overload and ecological deterioration that has come to typify this region.

Stress on local resources forced by rising population has been estimated quantitatively. Research by Yang Dingguo (1996) indicates that the population in Ganzhi and Arba in 1993 (1,665,000) exceeded the population that could be sustained by local energy provision by 392,000, local production of protein by 395,000 and local production of fat by 80,000. In 2000 the population rose to 1,923,000, the overloading index reaching 448,000, 415,000 and 32,000 respectively. Consequently, population growth in West Sichuan is the decisive root cause for ecological deterioration, although made manifest through over-pasturing and over-tilling.

### *Rural underdevelopment*

Related to rural underdevelopment, traditional agricultural practices employed by indigenous groups are thought to contribute to environmental overloading in three ways. First, settlement patterns. There is a tendency amongst indigenous groups to settle at increasing altitudes, thereby bringing a growing range of fragile habitats under population pressure. Movement up slope is a coping mechanism for indigenous groups seeking to avoid exposure to warfare or epidemic disease (Liu Huilan 1948). Second, nomadic herding. This is still a dominant livelihood form in West Sichuan and produces two ecological/economic problems. First, without an opportunity to improve grazing lands scope for increasing productivity of herds to ameliorate the negative impacts of population growth is limited. Also, transience excludes this group from using electricity as a form of energy supply. Manure is a principal source of fuel, with growing demand putting more pressure on grasslands both in terms of grazing extraction and in the loss of nutrients as manure is extracted. Third, resettlement. Since 1950, many communities have been resettled in the surroundings of the massive hydropower projects. The need to re-establish homes and land for their survival causes serious vegetal cover loss and soil erosion. This problem is repeated in many regions where big projects are located (Liu Xiao Quan *et al.* 2001).

### *Mode of production*

Inappropriate rural technologies have led to environmental degradation. A prime example is vertical tilling on hill slopes, which leads to water loss and soil erosion. As is shown in Table 14.4, in West Sichuan drought-prone land

*Table 14.4* Productive conditions of agriculture in West Sichuan, 1999

| *Prefecture* | *Proportion of drought land* | *15–25° slope* | | *Slope >25°* | |
|---|---|---|---|---|---|
| | | *Tilled land* | *Terrace to slope* | *Tilled land* | *Terrace to slope* |
| Ya'an | 61.03 | 18.39 | 18.49 | 22.08 | 8.02 |
| Ganzhi | 99.14 | 28.52 | 0.00 | 11.45 | 0.00 |
| Arba | 99.93 | 34.12 | 0.00 | 18.33 | 0.00 |
| Liangshan | 77.41 | 34.19 | 5.15 | 14.42 | 2.11 |
| Leshan | 44.33 | 18.61 | 16.46 | 10.03 | 8.75 |
| Panzhihua | 47.20 | 28.99 | 32.78 | 9.61 | 9.02 |

Note: All values are percentages.

is often tilled – for example, in Arba, Ganzhi and Liangshan some 34.12 per cent, 28.52 per cent and 34.19 per cent of land is tilled. Some 9–22 per cent of this land is on slopes of 25 degrees or more, with a further 18–34 per cent on slopes of between 15 and 25 degrees. Of dry land at a slope angle over 25 degrees almost 90 per cent is vertically tilled.

The economy and ecology of pastoral areas is caught in a cycle of poverty and environmental degradation as outlined in Figure 14.2. Most grassland is unimproved and productivity of natural grassland is very low, weakened as it is by pests. Low productivity, combined with the absence of the processing of livestock products, does not generate sufficient funds to upgrade pasture.

In forested areas, lumbering and lumber transportation dominate. Taking Ganzi prefecture as an example, in 1997 the economic value of lumbering and its transportation was over US$36 m. With a comparatively small timber products industry (economic output US$1 m) pressure is on extraction rather than value added production – consequently forest has vanished quickly in large areas. In this way, whilst only a limited economic return is enjoyed, large-scale environmental degradation occurs swiftly. Since the 1960s the flow of the upper Minjiang river in the dry season has been reducing gradually, with runoff variations increasing and the surrounding land becoming increasingly arid; all these local environmental changes can be linked to the reduction of forest in the river valleys.

Mining and metallurgy are the dominant industrial activities in the region, with small-scale enterprises making up 96.78 per cent of the sector. Small scale in this case has again prevented the application of modern environmentally friendly technology, leading to widespread water and air pollution and the degradation of the landscape. Gold exploitation in particular has led to the loss of good farmland and grassland.

### *Reconsiding exploitation and ecological deterioration in West Sichuan*

Both relatively and absolutely, West Sichuan is rich in land, forest, hydropower, water, minerals, medical herbs and tourist resources. But this has not ensured a sustainable relationship between ecology, economy and society. Growth in hazard risk and human vulnerability has resulted. What is lacking from recent regional development planning is appropriate and accessible science and technology and ecological consciousness. Although average population density in West Sichuan is high at 34.9 people per square kilometre, this is less than the mean figures for Sichuan or China; but overpopulation exists and intensive ecological destruction occurs. Why?

As has often been proposed, the population problem in China is a question of quality, not quantity. West Sichuan is characterized by poor education, economic under-development and serious ecological destruction. A lack of human resources and education has resulted in a deficiency of experts, leading to technical backwardness, a reliance on traditional production modes and related environmental impacts. This lack of human resources and skills, combined with policy inertia, is an obstacle for risk reduction.

In poor areas, intensive, pre-modern exploitation patterns are typical. Approaches to economic development in such circumstances usually rely on one of two basic strategies which I have termed 'exploitation profundity' and 'exploitation intensity'. Exploitation profundity is a lengthening of the local industrial production chain which should increase local economic benefit from resource exploitation. Exploitation intensity strategies increase the volume of resource being exploited without building value added into the local economy, thus building economic growth on the back of increased environmental pressure. To date, development in West Sichuan has been characterized by a strategy of intensity not profundity. This is typical of pressure-over-capacity and plundering-over-restoration strategies. It produces limited economic returns set against large-scale ecological destruction and resource exhaustion.

In an area dependent on primary resource extraction, industrial planning should be easy. But this is not the case in West Sichuan. There is no reason why all dominant resources need and ought to result in dominant industries. Some resources – forests being an example – should not be extensively exploited because of their ecological function. If a dominant resource is transformed into a dominant industry this process needs to be compatible with the environment. In West Sichuan the dominant industries of timber production, mining and agriculture have seriously spoiled the environment. At the same time, potentially less damaging industries such as tourism, medicinal plant extraction, organic food production and so on have not been established. Industrial development should be intensive in product refinement and development, research into new technology and market development, rather than being dependent entirely on the intensity of raw material extraction for growth. Therefore, both the right selection of industries and the promotion of profundity over intensity should be principles used for development policy in this region that can restore the deteriorating environment.

## *Strategic views*

Economic and ecological reconstruction in the upper reaches of the Yangtze valley require the strengthening of local capacity in new technology and greater access to information, human skills, the market and finance – all of which will need outside support, possibly international in character. A strategy of reconstruction should integrate elements of reforestation, the protection of remaining natural forest and grassland, the regeneration of degraded grasslands, new transport and marketing infrastructure, efforts to support local livelihood as a way to reduce economic poverty, ecological tourism, natural medicine exploitation, green farming, ecological energy and non-traditional forest goods with education, science and technology, community building, and the establishment of local markets as a mechanism for promoting a greater orientation towards a culture of competitive entrepreneurialism.

Moving towards a more harmonious relationship between the economy and the environment will require specific adjustments to the industrial and agricultural sectors. Industrial change will be centred upon the need to build up secondary and tertiary sector activity (Chen Guojie and Yang Dingguo 2000). Broadening regional industrial development in this way could reduce environmental stress caused by agriculture, whilst also stimulating new avenues for economic growth. Income derived from such new activities will then be available for investment in ecological reconstruction and conservation. How might secondary and tertiary industry be encouraged to invest in this region?

Three strategies seem most appropriate. First, West Sichuan is rich in natural and cultural resources that would be attractive to tourism development. Tourism could be oriented towards the national and international markets, both of which are likely to grow in the medium to long term. Of course, care must be taken to organize and regulate tourism so that it does not endanger the very environmental resources it should safeguard, as has often happened with tourist-led development elsewhere. Second, as part of an overall emphasis on 'green' industry, small mines exploiting a variety of minerals should be closed. The ecological impact of these works is out of proportion to the economic benefits they generate. This should not be seen as a presumption against small-scale business of any kind. Efforts should be made to develop alternative local income generators based, for example on natural pharmaceutical production, traditional Chinese medicine, non-traditional forest/grassland products (such as fibre, tannin, starch, perfume, drink, dye, paper pulp and forage), livestock rearing, ecological energy production, specialist and high-value agricultural products (such as pine down, Chinese prickly ash, apple, tea, mulberry, sugar, tobacco and high-land barley wine, etc.), craft work, and so on. Third, to replace unecological industry some of the areas of small-scale business identified above would also be suitable for larger-scale development. These might include natural pharmaceutical products, vegetables, flowers, organic food, high-quality agricultural products and fruits and dry fruits. These industries are agricultural in nature but offer higher returns; they have the added advantages of increasing plant cover in the region and fitting in with established livelihood strategies, so reducing the retraining needed by indigenous populations taking up work in these new sectors.

In addition to restructuring the productive sectors, internal changes in the practice of agriculture would help relieve environmental stress (Chen Guojie 1999b). Changes would focus on land use patterns. Old crop varieties and planting patterns have contributed to declining soil fertility and soil erosion; the adoption of new, appropriate crops and farming practices offers the potential opportunity for achieving new sources of agricultural income and alleviating soil erosion. This strategy has a number of components:

- First, an appropriate distribution of crop types is required. In the moderate and high mountain areas, grass and animal husbandry should dominate. Forestry and forest product extraction will dominate in low mountain areas. In lower, hilly land, sloping field cultivation should apply the techniques of terraced agriculture. In the river valley and flood plain more intensive production of cash crops, including cereals, would be preferred.
- Individual river catchments should form the basic planning unit for economic and ecological recovery.
- The encouragement of a movement from mixed subsistence livelihoods towards more specialized forms of production based upon the comparative advantages offered by each ecological zone.
- Further research into appropriate technology for grassland reconstruction and management.
- The research and promotion of mixed cropping systems designed to provide economic returns from cash crops, subsistence food needs and to act as soil cover to reduce soil loss. The Chengdu Institute of Biology at the Chinese Academy of Sciences has developed just such an agroforestry system, which has been piloted at an ecological observatory in Mao County. Apple trees are used to provide high cover, a cash crop – *dioscorea panthaica priain et burk* (a vine species, the raw material plant of Diao Pharmaceutical Company) is used as a mid-level shrub and vegetable, and legumes provide subsistence inputs and act as ground cover.

### *Reforestation of cultivated land*

Reforestation will include returning cultivated land to forest and banning further deforestation (Chen Guojie 2001). As with the policies outlined above these programmes aim to increase plant cover and so promote a good ecological cycle by the control of soil erosion. This will reduce the high variability of current surface runoff rates, allowing droughts, flooding and landslides to be mitigated at their environmental source. But this programme also addresses the economic roots of local environmental/ecological degradation by providing a pathway for local economic development. This is a timely period for reforestation and the banning of future deforestation, with many forest industry bureaux in the upper reaches of the Yangtze river having been forced to stop production because of the over-exploitation of forestry resources.

Within this broad framework local governments and planning authorities will have to base development policy on local environmental/ecological conditions. Some examples of the different environmental/ecological zones and preferred policy options are presented below:

1 Humid monsoon subtropical mountains and hills. These include the area to the east of Qinghai–Tibet plateau, Sichuan basin and its surrounding mountains, part of Yunnan–Guizhou plateau, part of Chongqing municipality and the upper reach of Hanjiang valley (an important branch of the Yangtze river). These areas favour afforestation; cultivation will need to be curtailed, with pioneer species such as grasses being planted first, followed by trees.

2 Karsts areas. Mainly found in the south of Sichuan Province, Chongqing municipality and Guizhou Province. Afforestation is very difficult here, but soil erosion is not serious. The cultivated lands in these areas are not intensive and tend to be widely dispersed. On hill slopes soil loss can be reduced by the promotion of terraced field agriculture.

3 The subalpine, forest-steppe zone. These areas are not suitable for cultivation; natural forest has typically been cleared some time ago, with landslides being a frequent consequence. The reintroduction of trees adapted to subalpine conditions should be the planning priority here.

4 Arid valleys. These include the Dadu, Jinsha, and upper reaches of the Mingjiang, Anling, and Longchuan rivers. Returning cultivated land to forest in these areas will be very difficult. Good irrigation systems will be a prerequisite for reforestation. But the cost of artificial irrigation is prohibitive at about US$3,658 per hectare. One way around this impasse may be to encourage the development of irrigation for industrial forestry over the short term in order to reclaim the costs of infrastructural investment and then to move on to large-scale afforestation.

5 The Qinghai–Tibet plateau This area includes the grassland in Ganzi and Aba prefectures in Sichuan Province and the source region of the Yangtze in Qinghai Province. This area is the most important ecological region in the upper reaches of the Yangtze river. Cultivation on steep hill slopes and the overgrazing of grasslands has led to serious environmental degradation. Returning cultivated lands to grass, and especially the construction of artificial grassland, will be the key to restoring environment here. More specific policies for consideration to facilitate this macroenvironmental change will include the settlement of nomadic herdsmen, the building of modern communities, the promotion of more ecologically sound energy sources to replace the tradition of burning animal dung, and the establishment of research and management agencies for the grassland.

### *Strategy for community construction*

Restructuring of the economic and ecological/environmental bases of this region is likely to impact greatly on indigenous groups, especially nomadic pasturalists. The following are policy options for coping with the social changes that will inevitably accompany such far-reaching policy. In the grassland regions new settlements will be needed to house previously nomadic groups. Such settlements should be planned under the guidance of eco-community theories that embody a harmony of population and environmental load. Basic facilities, including access to transportation, communication, education, medicine, scientific and technological services, and water provision, can be extended to previously nomadic groups in this way. Settlement of some groups in Ganzhi Tibetan Autonomous Region in Sichuan Province has already begun. Before 1995, most herdsmen were nomads; by the year 2000 some 36,000 people had settled down. This has required the construction of 3 million square metres of new houses, together with separate stables and improved grassland and fodder land for each family. Sedentary agriculture has now come to play a main part in the regional economy.

Improving local development opportunities is wrapped up in the need for local ecological and environmental reconstruction. This can include projects that contribute towards vegetation restoration, erosion control, water supply, energy substitutes and an increase in the accessibility of transportation. The Yaba village of Neiniaoqi County in Tibet is an example of a community that has benefited from this kind of environmental reconstruction. In 1993 the village had a population of 150 from 18 families and an annual income of 860 yuan. However, because of its location alongside Qu river, every five years about a third of the village was flooded and every 20 years it was entirely submerged. Since a reconstruction of the riverbeds in 1994 flooding has subsided and the local annual average income has increased to 1,200 yuan. Six new families have settled in the village. An alternative way around local environmental hazards is to relocate a community.

## Conclusion

The reconstruction of the economic and ecological systems of the upper reaches of the Yangtze river point to a number of areas for policy reform and where further research is required. These are set out below:

- A shift in the scale of planning is required from the current orientation grounded in centralized, state control towards local development planning. Strengthening local ecological and economic systems requires local knowledge that is hard to access and is in danger of being undervalued at the national level.
- A number of key technical problems need to be overcome. These include the controlling of rodent populations in grassland areas, how to restore forest in subalpine environments, how to restore the ecological diversity of deteriorated regions and how to improve the water-saving aspects of agriculture.
- Planners and farmers alike need to be convinced of the usefulness and realistic opportunities that the changes argued for in this chapter can bring for mitigating the effects of global environmental change in this sensitive region. Change may take some time. Some techniques like agro-forestry have already been experimented with and should be adopted

as soon and as widely as possible. Other techniques require further research and development.

- Technological weaknesses could be overcome by the development and introduction of plant strains that can return cultivated land to forest or grass cover.
- Entering the global market in ecological/agricultural products. There is an argument to be made for introducing exotic grass and tree species from other countries that can most rapidly commence the process of ground restoration. However, this must be offset by research into the modification of native species and the protection of local famous brands such as the Han Yuan prickly ash, Maoxian apple and Jinchuan pear. The potential international and national markets for these agricultural products should be exploited.
- Changes in planning should take into account the liberalization of the Chinese economy. To this end local and foreign direct investment should be encouraged. Local managers will need to be exposed to the new techniques that come with liberalization and business management or development planning in a market economy.
- Given both the region's sensitivity to global environmental change and the wide consequences across China of environmental changes in this region, regional ecological and environmental monitoring systems should be established. Monitoring should be able to supply information not only to prepare for natural disaster but also to maintain forest and grass ecosystems.

## References

Bao Weikai and Chen Qingheng (1999) 'Approaches about ecosystem recovery and reconstruction in deteriorated mountain area', *Journal of Mountain Science* 1, 17–22.

Chen Guojie (1999a) 'Industrial developmental issues in West Sichuan', *Territorial Economy* 4, 19–21.

Chen Guojie (1999b) 'On the environmental protection and direction of its industrial development in mountain areas', *Science and Technology Review* 2, 49–52.

Chen Guojie (2000) 'Viewpoints on ecological construction on upper reaches of the Yangtze river', *Journal of Science and Technology* 7, 59–61.

Chen Guojie (2001) 'Problems and countermeasures in returning cultivated land in steep hills into forest and in banning deforestation in the upper reaches of the Yangtze river', *Resources and Environment in the Yangtze River* 10 (6), 544–549.

Chen Guojie and Yang Dingguo (2000) *Comprehensive Exploration and Sustainable Development On the Mountainous Poor Connection Area of Chongqing Municipality, Hubei Province, Hunan Province and Guizhou Province*, Sichuan: Sichuan Science and Technology Publishing House.

Fan Yiping and Chen Guojie (2001) 'A study on qualification of function and measurement of sustainable development in Chengdu', *Sichuan Environment* 20 (1), 1–5.

Liu Huilan (1948) 'Alpine community in West Sichuan', *Acta Geographica Sinica* 15, 27–29.

Liu Xiao Quan, Chen Guojie and Chen Zhijian (2001) 'Ecological and environmental warning on rural habitat ecosystem – a case study of group 5 of Cizhu village in Wanxian city', *Acta Ecological Sinica* 21 (2), 295–301.

Peng Puchu and Fu Chen (1999) 'Progress of study on mountain vertical natural zones in China', *Geographical Sciences* 4, 303–308.

Sun Guanyou (1998) *The Mire and Peat Land of the Hengduan Mountain Region*, Beijing: Science Press.

Tang Bangxin (1995) *Natural Hazards and their Alleviation in Sichuan Province*, Changdu: The Press House of Electronic University.

Yang Dingguo (1996) 'A discussion on the main system of ensured projects for sustainable development along the upper Yangtze river valley', *Resources and Environment in the Yangtze Basin* 5 (2), 104–111.

# Part V

# Conclusion

# 15 Emerging concerns

*Mark Pelling*

## Introduction

This conclusion has two tasks. First, to summarize the principal findings of the volume's chapters. Second, to draw out those overarching issues raised by the contributors that might point the way towards future research and policy agendas.

## Disaster amidst 'glocalization'

Swyngedouw (1997) argued that contemporary development processes occur simultaneously at global and local scales, and that it is the interaction of global flows with local context that shapes history. For disaster mitigation and response the implication of this is that work must be undertaken at both the global and local levels and with intervening dynamic processes (Blaikie *et al.* 1994). By bringing together authors with expertise at the global and local scales, and with insights on the linkages that channel their interaction, this book has sought to make a contribution to debates on the interaction of the broad patterns of global socio-ecological change and the place- and time-bound status of specific disaster–development relationships. The chapters were arranged in three sections dealing in turn with global, international and local issues, and are summarized below.

The chapters in Part II (especially Wisner, and Adger and Brooks) made some progress in outlining the fundamental contours of the relationship between disaster risk and vulnerability as part of development patterns under the globalization of capital, Western-style democracy and global environmental change. Fordham, reminded us of the social diversity that should be considered, even when examining risk processes at a global scale, by critiquing the dominant orientation of disaster and development discourse from a feminist perspective. Etkin and Dore applied this basic knowledge and asked how it might be possible to identify the capacity of societies to cope with environmental risk in this era of rapid global change.

In Part III, contributors moved from an examination of global processes and trends to looking at the actors involved in shaping international dialogue and policy on environmental risk and vulnerability. Christoplos identified a range of actors from the public, private and civil sectors acting at local, national and international scales with stakes in disaster mitigation and response. Despite the variety of actors and their range of comparative advantages there remained scope for greater cooperation across sectors and scales. Kelman focused on the role played by natural disaster events in international relations, and found that under certain conditions disasters have in the past produced a softening of tensions between otherwise hostile nation-states. Whether disaster diplomacy could, or indeed should be used more mechanically as a diplomatic tool remains an open question. The spreading of disaster impacts through insurance, and the international re-insurance market was examined by Salt. He found an alarming degree of complacency within the sector, which remains liable for massive claims should major urban centres be hit by natural disaster. Collapse of the international insurance industry would have serious repercussions for the health of the global economy.

Whilst global pressures need to be identified and tracked, it is their interaction with local-level processes, policies and action situated in a specific historical context that shapes experiences of vulnerability and hazard. Part IV provided a number of innovative approaches in the study of disaster and development at the local scale. Homan extended a post-structural approach derived from within the political ecology canon, making clear the great extent to which the social construction of environmental risk and disaster differs over space, and that despite cultural globalization world-views remain tied to place and time. Morris took up certain elements of the post-structural methodology but argued that the physical nature of disasters and the urgency for material responses to mitigate risk limits the appropriateness of this approach. Allen problematized the contrasting conceptualizations of vulnerability held by partners in a community-based vulnerability reduction project facilitated by the Philippine National Red Cross – in this instance lack of shared understanding undermined local adaptive capacity. Warner drew on institutional analysis and management theory to reveal the interplay of individual agency and organizational structure in restructuring institutions for risk management. Özerdem focused on failings in the construction industry and broader weaknesses in government oversight that led to the huge losses in the Marmara earthquake, Turkey. He also provided an example where popular anger at government failure has produced some movement towards national reform. Given the rising profile of urban areas as sites of disaster (see Pelling 2002a, 2002b) this insight is particularly relevant to emerging patterns of risk globally. Guojie presented a review of the causes and possible policy responses to environmental degradation and disaster risk in the upper reaches of the Yangtze river. Guojie proposed radical restructuring of the industrial base and settlement patterns for this region. This raises questions over the distribution of costs and benefits, where present indigenous populations will bear the brunt of reform costs, with benefits spread amongst the wider Chinese and global present and future populations. Given the opening of the Chinese economy and its recent membership of the WTO, processes of socio-ecological change are set to become less easy for planners to influence.

## Where do we go from here?

Four areas of concern emerge from the chapters of this volume:

- a human rights perspective on disaster;
- the importance of scale;
- the necessity of integrating social and physical sciences;
- the need to find a balance between mitigation and adaptation.

These new directions of research and policy formation are connected in many ways, as the following discussion makes clear. At a fundamental level all have arisen in a particular historical context. They are set against a background of continuing and perhaps worsening losses to disaster with a natural trigger, despite 30 years of international disaster assistance (Pelling 2001). In the new millennium local, national and international systems of governance are in flux, pressured by global economic restructuring, post cold war politics, new information technologies, rural–urban and international migration and post-modern culture (see Chapter 1 and Mitchell 2001). To this we could add the influence of 11 September on US international policy, which has become increasingly focused on short-term fixes to address deep-rooted problems of international inequality. This limited outlook mirrors global trends in budgets for international disaster assistance, which have been increasing at the expense of commitment to the longer-term agenda of vulnerability reduction as part of ongoing development assistance (IFRC/RC 1998). Another spur for rethinking disaster mitigation and response has been the shortcomings of the UN International Decade for Natural Disaster Reduction (1990–1999) which, despite making progress in some areas of skill transfer and international networking (Pelling and Uitto 2001), failed to engage fully with the social, political and economic contexts of risk and the interplay of disaster with international development. In the following sections each of the four emerging concerns for disaster and development are elaborated by drawing on the academic literature to provide bases for further research ideas

### *Human rights*

> Everyone has the right to life, liberty, and security of person.
> (Universal Declaration of Human Rights, Article 3)

> Everyone has the right to a standard of living adequate for the health and well-being of himself [*sic*] and of his family . . . in the event of

> unemployment, sickness, disability, widowhood, or old age or other lack of livelihood in circumstances beyond his [*sic*] control.
>
> (Universal Declaration of Human Rights, Article 25)

The Universal Declaration of Human Rights (http://www.fourmilab.ch/etexts/www/un/udhr.html) already supports the right to personal security and for a basic standard of living during periods of unforeseen livelihood disruption. Amongst disasters analysts, there is a rapidly growing consensus in support of a human-rights-based approach to disaster assistance and mitigation – see the RADIX website (www/anglia.ac.uk/geography/radix), and a special edition of *Environmental Hazards* (2001). The human rights approach underlies the basic stance of *Natural Disasters and Development in a Globalizing World*, contributors share an orientation towards the rights (and responsibilities) of the individual, with nods to broadly communitarian political philosophy. The human rights agenda comes out most clearly from Wisner's (Chapter 3) push for reform in the UN system and Fordham's (Chapter 4) call for the rights of women and children, as well as their needs, to be given a central place in disaster policy and practice; it can also be seen, for example, to influence Allen's (Chapter 11) support for grassroots participation and Özerdem's (Chapter 13) support for popular protest following the Marmara earthquake. Chen Guojie's (Chapter 14) description of disaster–development from the top-down perspective of policy planning in the upper Yangtze river valley shows how difficult it is in practice to match a desire to enhance individual and collective claims for security with parallel claims on livelihood or cultural integrity, and similarly for planners to make strategic decisions that effect the individual rights of current generations against those of future generations.

The human rights agenda in disaster–development discourse is strategically targeted at modification in the international system (Mitchell 2001). As Wisner (2001) argues, the international community, and in particular the UN agencies, has provided technical knowledge, support for institution building and financial assistance for disaster mitigation and relief, but the 'missing ingredient is the kind of moral imperative that can mobilize local political will. It is when the world at large agrees to standards of responsibility by nation-states toward their citizens in the form of treaties, covenants and other agreements, that this moral force is felt most strongly' (2001: 126). Wisner proposes a human rights agenda catalysed at the international level by an Intergovernmental Panel on Natural Disaster run in parallel with the Intergovernmental Panel on Climate Change (IPCC) and given the remit of synthesizing technical knowledge and facilitating its application world-wide.

One approach for the human rights agenda might be to look for existing international codes of conduct in related areas of environmental management, industrial regulation and social development. An example of building an argument for extending rights already existing in parallel sectors of law would be to invoke the precepts of the precautionary principle. This is already recognized in international environmental pollution agreements and has acted to shift the point of government intervention from responding once damage has been inflicted to a more proactive stance on industrial pollution prevention. Importantly for disaster–development policy the precautionary principle presumes governments will act even when scientific evidence for risk is incomplete. Often a substance's interaction with other chemicals and its cumulative impact over the long-term is unknown, but government remains obliged to intervene in the interests of environmental security. By analogy this could be applied to the disaster–development relationship to provoke governments to increase investment in vulnerability reduction measures in the face of an unknown but probable future disaster risk.

The international consensus needed to push the rights agenda forward will take time, and requires new international institutions to be created. The long-term objective would be to empower citizens to challenge their own or other governments that do not make efforts to fulfil Articles 3 and 25 of the Universal Declaration of Human Rights in relation to natural disasters. This would act as a stick to provoke governments into taking more proactive steps to reduce disaster vulnerability in a coherent manner and by following internationally recognized guidelines for good practice. Kent (2001) sees civil society actors as being critical in pushing this agenda forward, perhaps by formulating a draft International Treaty on Disaster Mitigation and Relief; whilst the human rights agenda is being shaped there is clear role here too for empirical and theoretical contributions from disasters researchers.

### *Scale*

Issues of scale cut across the chapters in *Natural Disasters and Development in a Globalizing World*. Chapters included here, and the wider literature, suggest that scale interacts with disaster mitigation policy in two ways. First, in boundary mismatch (Cash and Moser 2000) where 'the boundaries of property and government, like the less sharply etched patterns of markets, rarely follow the outlines of biology and topography' (National Research Council 1996: 326). Here, for example, misfit between ecologically defined or risk zones and census or administrative districts leads to data-set incompatibility hindering policy development. A similar mismatch occurs when trying to synthesize processes working on human, ecological and geological timescales. Second, in the 'relitivization of scale' (Yeung 1998), different scales of institution – local, national and global – acting in concert in response to a common threat are often hampered by communication problems, inequities in decision-making power and competing interests where actions to reduce risk at one scale may increase risk at another.

Scale was identified as a critical issue in the production of disaster by Blaikie *et al.* (1994) who emphasized the importance of context and scale in their Pressure and Release Model. Here disaster is the outcome of two colliding pressures, one of human vulnerability and the other of environmental hazard. These pressures, but especially human vulnerability, can be traced back from immediately dangerous conditions (lack of basic services or secure housing or the absence of local institutions) to global and historical root causes tied to the international political economy. Between these two scales operate intervening dynamic pressures that manifest for example as migration flows and urbanization, flows of investment and unequal economic growth or political regime change and democratization. Throughout this book contributing authors have been at pains to identify such dynamic pressures and the pathways through which global pressures interact with national and local contexts to produce dangerous local conditions.

Moving forward an agenda for risk reduction requires not only research that links different scales in seeking explanations for disaster vulnerability but also an examination of the ways in which institutions function and interact at various levels in making decisions about the way disaster risk is perceived and responded to. Examples abound in the literature of local initiatives being constrained by lack of national or international support, particularly where grassroots organizations have to amend their development priorities to fit with the funding agendas of national or international agencies, and similarly of international efforts having inefficient or inequitable local impacts – for example, where relief aid is diverted by national or local level elites (see the chapters by Allen and Warner). The reasons for such failings are contextual, complex and politically charged, steeped as they are in concerns of sovereignty, political authority and legitimacy. Calls for enhanced action at the global level, for example in proposals for an Intergovernmental Panel on Disaster Reduction, need to be cognizant of the problems posed by cross-scale institutional interactions. For example, Mehta *et al.* (1999) show how the promotion of global approaches to problems of global environmental change has in certain contexts contributed to the undermining of local resource users' control over their environments, with enhanced local vulnerability a result (Adger *et al.* 2002).

Whilst analysts and practitioners are calling for greater global co-ordination in disaster policy, there is a parallel movement towards enhancing the power of individuals and grassroots actors operating at local scales in risk reduction. These two dynamics are reconstructing the architecture of governance regimes for disaster prevention. For best effect we need to understand not only how actors operate at each level, but how local-level, bottom-up participatory approaches articulate with international and national top-down agendas. Existing research on participatory development suggests that information asymmetries, unequal access to financial and technical resources and accountability towards donors rather than local participants/beneficiaries are significant barriers to meaningful partnerships (Lister 2000). Drawing on Ostrom *et al.* (1999), in the context of environmental risk Adger *et al.* (2002: 13) argue that 'overcoming these obstacles requires innovative forms of communication, information and trust in order to make links between scales and to make broader applications generalisable and transferable from one context to another'.

### *Integrating knowledge*

A number of chapters in this volume have presented combined physical and social scientific analysis (for example, Dore and Etkin at the national and global scales, Chen Guojie at the regional scale, Morris at the local scale). Most writers on disaster–development agree that integrating scientific knowledge and methodologies is essential in moving forward understandings of socio-ecological systems, and in so doing identifying both the fault-lines of social and ecological vulnerability and also opportunities for adaptation. There is a long tradition of collaborative social and physical science research in disasters studies, but moving from the summation of parallel research to truly integrated interdisciplinarity is proving difficult. In recent years impetus for social/physical science integration at the international level has come from the global environmental change programmes of international scientific unions. The International Council for Science took a lead in this by convening interdisciplinary forums and briefing documentation for the World Summit on Sustainable Development, 2002.

There are a number of methodologies that have been developed by individual sciences or through multidisciplinary research projects that offer pathways for integration. Examples include the narrative accounts of environmental history (for example: Arnold 1996; Gregory 2001), ecological economics modelling (see the journal *Ecological Economics*), and the use of integrated scenarios in assessments of institutional adaptability to socio-ecological change (Lorenzoni *et al.* 2000). Most familiar in studies of natural disaster and development, though, is the human ecology approach which sees people as constituent parts of co-evolving socio-ecological systems (see Blaikie *et al.* 1994; Varley 1994). Common ideas to social and physical sciences, such as resilience, adaptation, feedback and thresholds provide the basis of a common language for integration and have a good fit with social and physical science work in complexity theory. However, moving from theoretical constructions of integration to the collection of empirical material for verification is perhaps the biggest challenge.

A second and related barrier to be overcome is that between science and policy implementation. The outcomes of research need to be communicated to decision-makers in a clear and timely fashion. The experience of the IPCC is informative. Notwithstanding sociology of science criticisms of scientific neutrality (Haraway 1989), in a highly charged political arena the IPCC process has been seen to distance itself from political interference in its research to date, gaining credibility with governments, the public and other scientists and creating a relatively neutral political space for the generation of a global scientific consensus on climate change with which to inform international policy negotiation. The missing element in provoking policy change in the light of IPCC evidence is political will; this in turn stems from the perceived distribution of the costs and benefits to individual nation-states of adaptation or mitigation strategies (Mason 2001). In a similar fashion, the lack of political will in a context of competing policy priorities can also be seen to have limited the global spread of good practice generated by the UN RADIUS programme on urban earthquake mitigation techniques (see Wisner 2001). Understanding how scientific knowledge is translated into policy and how political imperatives shape science will be a necessary component of efforts to get beyond this impasse.

### *Mitigation versus adaptation*

Prevention is better than cure, but has proved to be illusive in disaster research and policy. Despite 30 years of technical research and a plethora of engineering-based solutions, joined more recently by efforts to reduce social vulnerability (see Allen, this volume), the disaster–development relationship remains dominated by disaster response. Recent work on adaptation (Adger 2001; Pelling 2002a, 2002b; Dore and Etkin, this volume) has sought to bring disaster response more fully into the development process by framing it as an opportunity for socio-ecological evolution. In theory, as well as being seen in the capacity of systems to cope with rapid change brought about for example by disaster events, appropriate adaptation has the potential to mitigate future risk as local socio-ecological systems components are brought away from thresholds of disaster risk (the prevention of mangrove loss or settlement of steep hill slopes for example). Under the globlization of disaster risk local adaptation is being framed as an alternative to global mitigation.

This can be seen most clearly in debates around global climate change. In this case the withdrawal of key actors (the US) has weakened global capacity for mitigating global climate change (with the principal global polluter outside of negotiations scope for reduction in greenhouse gas emissions is limited). The net result of reduced capacity for global mitigation is to push the onus for change away from the global and towards the local and national, and away from mitigation and towards adaptation. It is those facing the negative consequences of global environmental change that now have the responsibility to act. For those countries liable to high costs from mitigation and with capacity to adapt to global environmental change (mostly the richer countries of North America and Western Europe) this is good news; for those with low mitigation costs but at high risk from climate change and sea-level rise this is potentially catastrophic. For some, such as the Maldives, a global preference for adaptation may result in the first observed extinction of a nation-state through environmental change.

Citizens have the right to expect their governments to undertake reasonable efforts to mitigate disaster. Where this has clearly not been the case, as in the Marmara earthquake in Turkey (see Özerdem, this volume) the legitimacy of government will be called into question, and there is scope for adaptive reform if the government chooses to listen to its citizens. The thresholds needed for such political feedback loops to come into operation, and the extent of adaptive change they generate is the subject for further research. Between nation-states such feedback loops are not well developed. This partly explains Wisner's (2001) call for a universal human right to freedom from the unnecessary exposure to natural disaster risk. At present, at the international scale there is no obligation in international law to prevent states from undertaking activity that will cause environmental damage and potentially raise disaster risk in other states. Only on issues of industrial pollution and climate change is there momentum for reform. International negotiations rely on the voluntary willingness of states to accept a general obligation not to cause significant cross-border harm (Crawford 1999), based on a presumption of the equal moral standing of citizens of the home and other states. Where states place the interests of co-nationals above those of vulnerable 'outsiders' the moral equity argument collapses, and with it goes any scope for redress.

Linklater argues that '[t]ransnational harm provides one of the strongest reasons for widening the boundaries of moral and political communities to engage outsiders in dialogue about matters which affect their vital interests' (1998: 84). This supports the case, made by Wisner (2001) and his colleagues, for an international forum on natural disaster and development and a human right to be free from preventable harm caused by natural disaster. Mason (2001) argues that contemporary international dialogue on ozone depletion, climate change and Antarctica shows some evidence of fairness, with weak states receiving concessions; but there is clearly a long way to go before moral equivalence in the face of environmental risk is enjoyed internationally, and part of this journey must be squarely to face the distributional effects of policy preferences for adaptation over mitigation.

## Conclusion

Despite global flows of finance and information and resulting space–time compression, nation-states (national and local governments) remain the core actors in shaping disaster policy. Situated as they are at the juncture of local, national and international space, they are key in influencing the balance between mitigation, response or more integrated development strategies of adaptation. Decisions are made to varying degrees with regard to social equity (how are the costs and benefits of decisions to mitigate or adapt distributed in society?), economic efficiency (will resources have a greater impact if invested in disaster prevention or response or in more radical alternatives to build adaptive capacity?), and political legitimacy (do decisions reflect the public mood, do they add or detract from government authority in domestic and international spheres?). Strategic decision-making needs to be cognizant of local priorities, risks and scope for adaptation. Empowering local actors – community groups or local government – needs to be undertaken within a framework that links the local to the national and global. Providing institutional frameworks for timely information flow between international-level scientists and individuals having to cope with environmental risk amidst global change is perhaps the core challenge for natural disasters studies over the medium term. Disasters specialists are not alone in this endeavour and it dovetails into calls for revision of global/local systems of environmental governance and for reform of the relationship between political institutions and private sector actors (for example, in the WTO).

There is great scope for academic researchers, activists and policy-makers to feed into this ongoing process of disaster–development integration. The four themes identified in this concluding section offer potential starting points for academic contributions, with political ecology and complexity theory being well placed to provide theoretical and methodological frameworks. Of course it is likely that there are many more ways in which analyses of social and environmental change in the context of global flows and local contexts can inform disaster studies, and we hope that this book might at least contribute to debate from which new directions of analysis and policy will emerge.

## References

Adger, W.N. (2001) 'Scales of governance and environmental justice for adaptation and mitigation of climate change', *Journal of International Development* 13, 921–931.

Adger, W.N., Brown, K., Fairbrass, J., Jordan, A., Paavola, J., Rosendo, S. and Seyfang, G. (2002) *Governance for Sustainability: Towards a 'Thick' Understanding of Environmental Decision Making*, CSERGE Working Paper No. EDM 02–04, University of East Anglia.

Arnold, D. (1996) *The Problem of Nature: Environment, Culture and European Expansion*, Oxford: Blackwell.

Blaikie, P., Cannon, T., Davis, I. and Wisner, B. (1994) *At Risk: Natural Hazards, People's Vulnerability and Disasters*, London: Routledge.

Cash, D.W. and Moser, S.C. (2000) 'Linking global and local scales: designing dynamic assessment and management processes', *Global Environmental Change* 10, 109–120.

Crawford, J. (1999) 'Second report on state responsibility', *UN General Assembly A/CN.4/498*, New York: United Nations.

Gregory, D. (2001) *The Colonial Present*, Oxford: Blackwell.

Haraway, D. (1989) *Primitive Visions: Gender, Race and Nature in the World of Modern Science*, London: Routledge.

International Federation of the Red Cross and Red Crescent (IFRC/RC) (1998) *World Disasters Report 1998*, Oxford: Oxford University Press.

Kent, G. (2001) 'The human right to disaster mitigation and relief', *Environmental Hazards* 3 (3–4), 137–138.

Linklater, A. (1998) *The Transformation of Political Community: Ethical Foundations of the Post-Westphalian Era*, Cambridge: Polity Press.

Lister, S. (2000) 'Power in partnership? An analysis of an NGO's relationships with its partners', *Journal of International Development* 12, 227–239.

Lorenzoni, I., Jordan, A., Hulme, M., Turner, R.K. and O'Riordan, T. (2000) 'A co-evolutionary approach to climate change impact assessment: Part I. Integrating socio-economic and climate change scenarios', *Global Environmental Change* 10, 57–68.

Mason, M. (2001) 'Transnational environmental obligations: locating new spaces of accountability in a post-Westphalian global order', *Transactions, Institute of Britsh Geographers* 26, 407–429.

Mehta, L., Leach, M., Newall, P., Scoones, I., Sivaramakrishnan, K. and Way, S.-A. (1999) *Exploring Understandings of Institutions and Uncertainty: New Directions in Natural Resource Management*, Discussion Paper No. 372, Institute of Development Studies, University of Sussex.

Mitchell, J.K. (2001) 'Policy forum: human rights to disaster assistance and mitigation', *Environmental Hazards* 3 (3–4), 123–124.

National Research Council (1996) *Colleges of Agriculture at the Land Grant Universities: Public Service and Public Policy*, Washington, D.C.: National Academy Press.

Ostrom, E., Burger, J., Field, C.B., Norgaard, R.B. and Policansky, D. (1999) 'Revisiting the commons: local lessons, global challenges', *Science* 284, 278–282.

Pelling, M. (2001) 'Natural disaster?', in N. Castree and B. Braun (eds) *Social Nature: Theory, Practice and Politics*, Oxford: Blackwell.

Pelling, M. (2002a) 'Assessing urban vulnerability and social adaptation to risk: evidence from Santo Domingo', *International Development Planning Review* 24 (1), 59–76.

Pelling, M. (2002b) *The Vulnerability of Cities: Social Adaptation and Natural Disaster*, London: Earthscan.

Pelling, M. and Uitto, J. (2001) 'Small island developing states: natural disaster vulnerability and global change', *Environmental Hazards* 3, 49–62.

Swyngedouw, E.A. (1997) 'Neither global nor local: "glocalization" and the politics of scale', in K. Cox (ed.) *Spaces of Globalization: Reasserting the Power of the Local*, New York: Guilford.

Varley, A. (ed.) (1994) *Disasters, Development and Environment*, New York: John Wiley.

Wisner, B. (2001) 'Disasters: what the United Nations and its world can do', *Environmental Hazards* 3 (3–4), 125–127.

Yeung, H.W.C. (1998) 'Capital, state and space: contesting the borderless world', *Transactions, Institute of British Geographers* 23, 291–309.

# Index